Palaeontology —
An Introduction

Related Pergamon Titles of Interest

Books

ALLUM:
Photogeology and Regional Mapping

ANDERSON:
Field Geology in the British Isles (A Guide to Regional Excursions)
The Structure of Western Europe

ANDERSON & OWEN:
The Structure of the British Isles, 2nd edition

BOWEN:
Quaternary Geology

CONDIE:
Plate Tectonics and Crustal Evolution, 2nd edition

EYLES:
Glacial Geology

FERGUSSON:
Inorganic Chemistry and the Earth (Chemical resources, their extraction, use and environmental impact)

*HALLAM:
The Palaeontological Microfiche Library

HENDERSON:
Inorganic Geochemistry

OWEN:
The Geological Evolution of the British Isles

ROBERTS:
Introduction to Geological Maps and Structures

SIMPSON:
Geological Maps
Rocks and Minerals

Journals

Computers & Geosciences

Journal of African Earth Sciences

Journal of Structural Geology

Quaternary Science Reviews

Inspection copies of any of the above books (except where indicated by an asterisk*) available to academic staff, without obligation, for their consideration for course adoption or recommendation. Full details of this service and of all Pergamon books and journals and a free specimen copy of any Pergamon journal available on request from your nearest Pergamon office.

*Not available under the terms of the inspection copy service

Palaeontology —
An Introduction

by

E. W. NIELD
formerly Poroperm Laboratories, Chester, UK

and

V. C. T. TUCKER
formerly Yale Sixth Form College, Wrexham, Clwyd, UK

PERGAMON PRESS
OXFORD · NEW YORK · TORONTO · SYDNEY · FRANKFURT

U.K.	Pergamon Press Ltd., Headington Hill Hall, Oxford OX3 0BW, England
U.S.A.	Pergamon Press Inc., Maxwell House, Fairview Park, Elmsford, New York 10523, U.S.A.
CANADA	Pergamon Press Canada Ltd., Suite 104, 150 Consumers Road, Willowdale, Ontario M2J 1P9, Canada
AUSTRALIA	Pergamon Press (Aust.) Pty. Ltd., P.O. Box 544, Potts Point, N.S.W. 2011, Australia
FEDERAL REPUBLIC OF GERMANY	Pergamon Press GmbH, Hammerweg 6, D-6242 Kronberg-Taunus, Federal Republic of Germany

Copyright © 1985 E. W. Nield and V. C. T. Tucker

All Rights Reserved. No part of this publication may be reproduced, stored in a retrieval system or transmitted in any form or by any means: electronic, electrostatic, magnetic tape, mechanical, photocopying, recording or otherwise, without permission in writing from the publishers.

First edition 1985

Library of Congress Cataloging in Publication Data
Nield, E. W.
Palaeontology, an introduction.
Includes index.
1. Paleontology.   2. Paleobotany.
I. Tucker, V. C. T.   II. Title.
QE711.2.N54 1985   560   84-26515

British Library Cataloguing in Publication Data
Nield, E. W.
Palaeontology: an introduction.
1. Paleontology
I. Title   II. Tucker, V. C. T.
560   QE711.2

ISBN 0-08-023854-8 Hardcover
ISBN 0-08-023853-X Flexicover

Printed in Great Britain by A. Wheaton & Co. Ltd, Exeter

Acknowledgements

Both authors wish to express their sincere gratitude to Professors T. R. Owen and D. V. Ager (University College, Swansea) for their encouragement and support in writing this book, and for ensuring a continuity of teaching which extends between (and beyond) those widely separated periods when each author attended the Swansea department as an undergraduate. We extend our thanks to the Principal and Governors of Yale Sixth Form College, Wrexham, and the geology departments of the National Museum of Wales, Cardiff, and Swansea University for access to their fossil collections. Finally, thanks are due to our families and friends for their continuing help and boundless patience.

EWN
VCTT

About This Book

This book is designed for students embarking upon a study of palaeontology. It highlights the importance of scaled, annotated drawings and concentrates upon those features of greatest use in the identification of forms likely to be encountered in early laboratory studies and fieldwork.

Each chapter on the major invertebrate fossil groups (Chs. 3–8) has two sets of illustrations. The first set is intended not only to support the text, but to act as the principal source of much factual material. Our intention has been to reserve text space for more important (and interesting) subjects than the precise expression in words of the simpler morphological features and measurable parameters. For this reason, readers are strongly advised to examine these figures quite as attentively as the written pages for descriptive terms and their meanings.

The second set of illustrations in these chapters consists in each case of scaled, annotated line drawings of named specimens, put forward as examples to students of the kind of work they should be producing in practicals. But in as much as they represent a wider section of the whole range of form present in any group, they may also be seen as additional sources of morphological information. They should be carefully examined — and even copied, in the absence of suitable specimens. No adequate study of palaeontology can be made without a secure knowledge of its vocabulary.

Nevertheless, the book is intended as more than a manual of features and a dictionary of their scientific names. We have tried to give a picture of a broadly based natural science, full of questions, rich in new approaches to the understanding of nature's history. We have also tried to give some idea of how the science actually works, and of how it regulates itself.

So — notwithstanding the great importance of drawing and precise description — we hope to impress upon readers that these are only the groundwork for the real business of palaeontology; the means, not the end. To aid correlation, investigate evolution and to reconstruct the modes of life and interactions of ancient animals in vanished ecosystems — these are the legitimate aims of our endeavours.

By *understanding* (not just recognizing and naming) fossil forms, we can analyse their function; by deducing function we illuminate evolution, and by appreciating that fundamental principle of nature, we provide more reliable correlation and are able to see, a little more clearly, the capricious process which has led to our own existence. This science, like any other, is less about facts than the ideas which bind the facts together.

EWN
VCTT

Contents

List of Tables   xi

1. **Introduction — Fundamental Questions**   1
    - 1.1 What is palaeontology?   1
    - 1.2 What is a fossil?   1
    - 1.3 What features and conditions are favourable for preservation?   1
    - 1.4 Of what materials are 'hard parts' composed?   2
    - 1.5 Can a complete organism become fossilized?   3
    - 1.6 What are the main changes during fossilization?   3
    - 1.7 What are moulds and casts?   4
    - 1.8 How do fossils 'date' rocks?   4
    - 1.9 What are fossils 'occurring *in situ*'?   5
    - 1.10 What are 'derived' fossils?   6
    - 1.11 How may fossils be used as indicators of environment?   6
    - 1.12 What is taxonomy?   8
    - 1.13 Who creates the classification?   9
    - 1.14 What is 'numerical taxonomy'?   10
    - 1.15 Welcome to palaeontology   11

2. **The Origin of Life and the Earliest Fossils**   12
    - 2.1 When and how might life have begun?   12
    - 2.2 What can geology tell us about this early life?   13
    - 2.3 Precambrian animals   13
    - 2.4 The Phanerozoic begins   14

3. **Trilobites**   16
    - 3.1 Introduction   16
    - 3.2 Ecdysis   16
    - 3.3 Morphology   16
        - 3.3.1 Facial sutures   18
        - 3.3.2 Eyes   18
        - 3.3.3 Limbs   19
    - 3.4 The nature of the trilobite cuticle   20
    - 3.5 Life cycle and development   21
    - 3.6 Classification   22
    - 3.7 Mode of life   23
    - 3.8 Geological history and stratigraphic value   26

4. **Graptolites**   30
    - 4.1 Introduction   30
    - 4.2 Morphology   30
        - 4.2.1 Dendroids   30
        - 4.2.2 Graptoloids   32
    - 4.3 Classification of the graptoloids   33
    - 4.4 The structure of the graptolite periderm   34

    4.5 Graptolite affinities. What was the graptolite animal? 35
    4.6 Mode of life 36
    4.7 Geological history and stratigraphic value 37

5. Brachiopods 44

    5.1 Introduction 44
    5.2 Morphology and internal anatomy 44
        5.2.1 Soft parts 45
        5.2.2 Hard parts (articulates) 47
        5.2.3 Hard parts (inarticulates) 49
    5.3 Shell growth, composition and structure 50
    5.4 Design and mechanics of articulate hinges 52
    5.5 Mode of life 53
    5.6 Classification 55
    5.7 Geological history and stratigraphic value 60

6. Molluscs 64

    6.1 Introduction 64
    6.2 General molluscan anatomy 64
    6.3 A diversity of molluscs 65
    6.4 Growth in coiled shells 66
        I. GASTROPODS 66
    6.5 Soft parts and life cycle 67
    6.6 Shell form 69
    6.7 The anatomical consequences of torsion 69
    6.8 Archaeogastropods 69
    6.9 Caenogastropods 69
    6.10 Pulmonates 71
    6.11 Mode of life 72
    6.12 Evolutionary history and geological value 72
        II. CEPHALOPODS 74
    6.13 Soft parts and life cycle 74
    6.14 Shell form 75
        6.14.1 Cameral deposits 75
        6.14.2 Coiling 76
        6.14.3 Shell aperture 77
        6.14.4 Opercula 77
        6.14.5 Ornament 78
    6.15 Belemnites 79
    6.16 Ammonites 79
    6.17 Mode of life 79
    6.18 Sexual dimorphism 82
    6.19 Geological history and stratigraphic value 83
        III. BIVALVES 84
    6.20 Soft parts and life cycle 84
    6.21 Shell form 88
    6.22 Mode of life 93
        6.22.1 Burrowing and boring 93
        6.22.2 Byssal tethering and free living 96
        6.22.3 Cementation 98
    6.23 Geological history and stratigraphic value 98

7. Echinoderms 99

    7.1 Introduction 99
        I. ECHINOZOA — CLASS ECHINOIDEA 99
    7.2 Shell form and soft parts of a regular echinoid 99
    7.3 Irregular echinoids 101
    7.4 Mode of life 101
    7.5 Spired echinoids, sand dollars and echinoid infaunalism 102
    7.6 Geological history and stratigraphic value 105
        II. CRINOZOA — CLASS CRINOIDEA 106
    7.7 Skeletal construction 106
    7.8 Mode of life 107
    7.9 Geological history and stratigraphic value 108

8. Corals and Stromatoporoids — 110

I. CORALS
8.1 The coral animal — 110
8.2 The coral skeleton — 110
8.3 The colonial habit — 110
8.4 Rugose corals — 111
8.5 Tabulate corals — 113
8.6 Scleractinian corals — 114
8.7 Geological history and stratigraphic value — 116

II. STROMATOPOROIDS

9. Vertebrates — 121

9.1 Introduction — 121
9.2 The vertebrate plan — 121
9.3 Jawless fishes — 122
9.4 The origin of jaws and lateral fins — 122
9.5 The origin of bone and lungs — 123
9.6 Amphibians and terrestrial vertebrates — 125
9.7 Goodbye to the egg — the origin of mammals — 127
9.8 The process of man's emergence — 129

10. Microfossils — 130

10.1 When is a fossil a microfossil? — 130
10.2 The unicellular organism and the five kingdoms — 130
10.3 Size and scale in single cells — 131
10.4 Kingdom Monera — 131
    10.4.1 Blue-green algae — 131
10.5 Kingdom Protista — 132
    10.5.1 Dinoflagellates — 132
    10.5.2 Diatoms — 132
    10.5.3 Coccoliths — 133
    10.5.4 Radiolaria — 134
    10.5.5 Foraminifera — 135
10.6 Kingdom Planta — spores and pollen — 135
    10.6.1 The alternation of generations and the rise of land plants — 135
    10.6.2 Typical spore and pollen morphology — 138
    10.6.3 Geological and stratigraphic value — 140
10.7 Kingdom Animalia — 140
    10.7.1 Ostracods — 140
    10.7.2 Conodonts — 142

11. Vascular Plants — 144

11.1 Introduction — 144
11.2 Palaeozoic plants without seeds — 144
    11.2.1 'Psilophytes' — 144
    11.2.2 Lycopsida — 144
    11.2.3 Sphenopsida — 147
    11.2.4 The origin of true leaves — 148
    11.2.5 The 'fern' complex — 148
    11.2.6 Filicopsida — 148
    11.2.7 Plant evolution reflected in Devonian spore assemblages — 148
11.3 Seed plants of the pre-Permian — 148
    11.3.1 Pteridospermales — 148
    11.3.2 Cordaitales — 148
    11.3.3 Coniferales — 150
11.4 Non-flowering plants of the Permian and post-Permian — 150
    11.4.1 Ginkgoales — 150
    11.4.2 Cycadales and Bennettitales — 151
11.5 Flowering plants — 152
    11.5.1 Angiospermopsida — 152

12. The Meaning of Fossils — 154

    12.1 Design, function, purpose — 154
    12.2 The function of the ammonite septum — 155
    12.3 The persistent paradigm — 156
    12.4 Perfect and not so perfect — 156
    12.5 Variation and natural selection — 158
    12.6 Natural selection under changing conditions — 158
    12.7 Some evolutionary case histories — 159
        12.7.1 Industrial melanism in the peppered moth — 159
        12.7.2 Evolution in Cretaceous irregular echinoids — 159
        12.7.3 The evolution of horses — 161
    12.8 Evolutionary trends and 'orthogenetic' theories — 161
    12.9 Small steps and giant leaps — 162
    12.10 Extinction — 163

Suggested Further Reading — 164

Author Index — 165

Systematic Index — 166

General Index — 170

List of Tables

Table 1.1	Stratigraphic Reference Chart, with Approximate Dates	7
Table 3.1	Fossil Arthropods (excluding Trilobites)	16
Table 3.2	Trilobite Classification	24
Table 4.1	Nomenclature of Graptolite Rhabdosomes	33
Table 4.2	Graptoloid Classification	35
Table 5.1	Summary of Brachiopod Geological History	45
Table 5.2	The Classification of the Brachiopoda	59
Table 9.1	A Classification of the Vertebrates	121
Table 10.1	Size Ranges of Microfossil Groups	131
Table 11.1	Broad Pattern of Plant History	150

1.
Introduction — Fundamental Questions

1.1 What is palaeontology?

Palaeontology is the study of ancient life through its fossil remains or the traces of its activity as recorded by ancient sediments. It is important because life on this planet has not always been as it is now. By studying the fossils in progressively older rocks, the palaeontologist attempts to establish an account of how all the animals and plants which make up the modern biosphere evolved from their earliest beginnings.

Clearly, it is important that palaeontologists should be both geologists and biologists. Their geological training inclines them toward using fossils to correlate rocks and establish the relative ages of rock units, while their biological background inclines them to working out how these ancient animals and plants actually lived.

Neither task is very easy. The incompleteness of the fossil record may be sufficient to mask many of the finer gradations in the evolution of life. Or perhaps the rock types under study do not contain the vital fossils needed for relative dating. Fossilization tends to preserve only the hard parts of organisms, like bones, shells and teeth. What can these things possibly tell us about the way organisms lived? Moreover, there may be no living plants or animals even remotely like the fossil ones.

The story of palaeontology is the story of how these and other obstacles have been, and are being, overcome. Literally, 'palaeontology' means 'a discourse on ancient beings', but it has developed into a complex and exciting field which takes as its subject the biology, not just of one moment in geological time, but of the 3500 million years or so during which life has flourished on earth.

Palaeontology establishes the evolutionary development of life. By doing so, it helps the geologist to put his (or her) rocks in order. It seeks to account for not only the regular minor extinction of individual species, but also the periodic mass extinctions which have, in the past, wiped out whole sections of the living world. By combining with the interpretations of the sedimentologist, it re-creates the environments in which fossil organisms once lived; and by applying engineering principles and the laws of physics to fossil skeletons, it establishes the reasons behind the design of mysterious skeletal features. Palaeontology is simply the history of nature, and the natural scientific answer to the question 'Where do we come from?'

With advances in our understanding, many subdivisions have arisen within palaeontology. The study of fossil plants is known as **palaeobotany**, and while this would suggest the title 'palaeozoology' for the study of fossil animals, the term is not generally used. **Palaeobiology** tends to be used to refer to the more interpretative 'biological' aspects of the science, although these are perhaps more correctly combined with the study of fossil communities under the general title of **palaeoecology**. The study of small fossils for which a microscope is often (though not always) needed is known as **micropalaeontology**. This science has a special botanical branch, concerned with fossil spores and pollen, called **palynology**.

1.2 What is a fossil?

Literally, 'fossil' means 'that which is dug up' — but the modern meaning of the word has acquired many refinements. For something to be a fossil, it must either be the remains of an ancient organism, or the trace of the activity of such an organism. But 'fossil' as an adjective may be used to refer to inorganic things; for example, a 'fossil volcano' or a 'fossil sand dune'. In these cases reference is made to their former existence before burial and preservation.

So there are two types of true fossil — **body fossils** and **trace fossils**. Body fossils are the actual remains of organisms. Trace fossils are indirect signs of life; dinosaur footprints, worm burrows, trilobite grazing trails and other evidences of life processes, such as fossil excrement (**coprolites**).

1.3 What features and conditions are favourable for preservation?

The soft, fleshy tissues of an organism may be

preserved in certain exceptional circumstances (see below), but the possession of hard parts vastly increases an animal's chances of being successfully fossilized. A jellyfish, for example, is far less likely to form a fossil than, say, a sea urchin.

But even hard parts are not indestructible, and need to be buried fairly quickly to prevent damage. Rapid sedimentation therefore encourages good preservation. Fine-grained sediments are also good for preserving fossils, on account of their low oxygen content and the fine detail which may be traced. And obviously, organisms living in water, especially sea water, always have the best preservation potential.

Fossils may occur in lake sediments, but then such sediments do not form a very significant proportion of the total geological record. Rivers tend to be rather vigorous and fluctuating for consistent preservation, but muds and silts associated with fluvial environments may well be fossiliferous.

Animals and plants living on land stand the poorest chances of preservation. Naturally, they also tend to occur in lake and river sediments, though they may also be found in some very unusual deposits, such as the tufa surrounding mineral springs, in volcanic sediments or in tar pits and peat bogs.

1.4 Of what materials are 'hard parts' composed?

Invertebrate animals (animals without backbones) may have durable external skeletons such as shells. And even soft-bodied invertebrates such as worms may have some resistant components (jaws, for example) which may be detected. Common, preservable skeletal substances include:

(a) *Silica* — SiO_2, silicon dioxide; a highly resistant material which forms the skeletal elements (**spicules**) of certain sponges.

(b) *Calcite* — $CaCO_3$, calcium carbonate; calcite is a stable crystal form (or 'polymorph') of calcium carbonate, and occurs in the skeletal plates of echinoderms and in many other organisms.

(c) *Aragonite* — $CaCO_3$, calcium carbonate; aragonite is less stable out of sea water than calcite, but it is a very common shell material. After burial, aragonite may change to calcite or be dissolved out and replaced by another mineral. Many molluscs have aragonitic shells.

(d) *Chitin* — Chitin is a polysaccharide — a complex, insoluble organic substance made of carbon, nitrogen, hydrogen and oxygen atoms joined in chains to form long molecules. Though not as durable as some mineral skeletons, it is commonly preserved. Insect exoskeletons are made of chitin. The trilobites possessed chitinous carapaces which were further strengthened by impregnation with mineral substances.

(e) *Scleroprotein* — Another group of complex substances, insoluble in water, which form tough coverings of certain animals. Substances such as *keratin* and *collagen* fall into this group. They are fibrous proteins, and they formed the skeletons of the graptolites. Molluscs also make use of a fibrous protein, known as *conchiolin*.

Vertebrate animals have internal skeletons composed of bone or cartilage or both. Cartilagenous skeletons, such as those of sharks, are rarely preserved. Bones and teeth, on the other hand, have a high preservation potential — especially teeth. Sharks' teeth are continuously produced in large numbers and so make excellent fossils. Bone is less durable, and is porous, consisting in life of cells, protein and a framework of mineral salts, notably calcium phosphate.

Whole vertebrate skeletons are rarely found intact because the bones are easily separated from each other after death. This has led to some difficulty in reconstructing these animals (especially so in the case of one eminent American palaeontologist, who placed a dinosaur's skull on the end of its tail). Disintegration is also a problem afflicting the study of fossil plants, which are nearly always found as fragments. It can take years of research to establish such basic facts as which leaf belonged to which stem, which seeds to which cones, which pollen to which flowers, and so on.

The preservation of plant material is a very chancy business. Everyone knows that coal is fossilized plant debris, but there is so little actual plant structure remaining in coal that it is only of limited use in palaeobotany. Useful plant fossils tend to be preserved in three basic ways: as **impressions**, **compressions** or **petrifactions**.

An impression contains no actual plant material. It is merely the form of, say, a leaf with its outline and its veins, impressed upon a bedding plane of very fine clay or silt. The famous clay pits at Puryear, Tennessee, yield Eocene plant remains which are preserved in this way.

Compressions, by contrast, preserve much of the original organic matter, usually as a black carbonaceous film. The preservation is not perfect, but may well display the more resistant parts of plant structure, such as the epidermal cells with their hard outer cuticle, spores, thorns and any thickened tracheids (veins) which may be present. The plants which are found around coal seams are usually compressions, as are the well-known Jurassic floras of Robin Hood's Bay, Yorkshire.

In the sections below, the processes of change which may transform original organic matter into other substances are described, but one deserves mention here as it is of especial importance to plant fossils. Plant fossils contain no mineral matter of any significance. Rapid burial and protection from the agents of decay are therefore of prime importance. And while these alone may be enough, the impregnation of the plant material with mineral salts may produce beautifully detailed fossils which can survive for hundreds of millions of years. Such impregnation may be brought about by carbonates of magnesium or calcium, by iron sulphide (pyrites) or silica. In Yellowstone National Park, in the slopes of the Big Horn Mountain, silicified tree trunks may be seen, weathered out of the volcanic sediments which buried them. A sequence of thirty fossil forests may be examined, one on top of the other, each representing the destruction of woodland following local volcanic eruptions 50 million years ago. Preservation of the wood is so perfect that even the filaments of parasitic fungi may be seen in the cells.

1.5 Can a complete organism become fossilized?

Yes, but only in very unusual circumstances. For example, natural refrigeration under glacial conditions in Siberia and Alaska has preserved the carcases of woolly mammoth and rhinoceros. Although around 45,000 years old, the meat of these animals has often been fit to eat, and their stomachs have even contained perfectly preserved undigested grass.

Extraordinary preservations of this kind often end up as celebrated museum exhibits. Amber is fossilized tree resin, and since it is very sticky when it first emerges, insects are often captured upon its surface and later entombed. Oligocene deposits along the Baltic coast have provided many spectacular examples, including one spider whose threads can still be seen as they emerge from its spinnerets.

Decay is seriously inhibited in peat bogs, and animals which have the misfortune to be trapped in them may be beautifully preserved. Irish peat bogs have yielded the giant Irish Elk, and some very famous human remains have been unearthed in Europe. The Tollund man, named after the fen where he was discovered, is perhaps the most spectacular example. When found, his leather cap was still secure on his perfectly preserved head. Every facial hair was visible, every pore and wrinkle of the skin. The noose around his neck also pointed to the non-accidental nature of his death 2000 years ago (Fig. 1.1).

In areas where natural petroleum seeps to the surface, tar pools or asphalt lakes may form. They are dangerous traps for animals of all kinds, and whole 'food chains' may be represented: the grazing animals, the predators which went after supposedly easy prey, and the scavengers which hoped to profit by their death.

FIGURE 1.1. An example of unusual preservation. This man was hanged, and his body thrown into Tollund Fen, Denmark, about 2000 years ago. Total exclusion of oxygen resulted in near perfect preservation. Note the leather cap, the rope, the delicate features of the skin and the bristles on the chin. Based on a photograph taken by L. Larsen, photographer of the Danmarks Nationalmuseet.

In very hot, dry regions, the corpses of animals may be dried so quickly as to prevent decay. In such cases the hide, as well as the skeleton, is preserved. This is known as mummification, and examples include certain giant sloths from New Mexico.

Although spectacular and of local importance, the number of animals fossilized in this way is insignificant. The mechanisms of common fossilization, while less spectacular, have yielded much the greater portion of palaeontological knowledge.

1.6 What are the main changes during fossilization?

Although the materials which compose animal skeletons are durable, it is very common for them to undergo some mineral transformations during the long processes of post-depositional change. These changes are related to those which turn soft sediment into hard rock, and are grouped together under the term 'diagenesis'. The changes in the fossils often reflect the diagenetic changes of their host rock.

(a) Recrystallization

It has been mentioned that many shells, particularly those of the molluscan groups such as gastropods and cephalopods, have aragonitic shells. Since aragonite is unstable out of sea water, it does not survive long after burial. Indeed, one can safely predict that no sedimentary rock older than the Mesozoic will contain aragonite, and indeed few rocks older than a few hundred thousand years still show unaltered aragonitic fossils.

Aragonite may invert to the stable polymorph of calcium carbonate, calcite. In such cases the original microstructure of the shell may be preserved in the new mineral (Fig. 1.2). More commonly, however, the aragonite will dissolve to leave a void.

(b) Replacement

The void left by the dissolution of an aragonitic shell is called a mould. This mould may be infilled at a later date by the precipitation of another mineral (Fig. 1.2). This may be calcite, but it could also be silica. If dissolution and replacement are separated by a void stage, all the original microstructure is lost.

Replacement by other minerals such as iron pyrites (iron sulphide, FeS_2), siderite (iron carbonate, $FeCO_3$), limonite (iron oxide, FeO) and haematite (iron oxide, Fe_2O_3) is commonly seen in iron-bearing rocks. Only pyrites forms very well-preserved fossils, however, and these must be protected from oxidation once they have been removed from the rock.

(c) Permineralization

This is the partial replacement or impregnation of original material by mineral salts, as described above in connexion with plant material. Bone, like wood, is highly porous and so is susceptible to this form of petrifaction.

1.7 What are moulds and casts?

When a foundryman casts a bronze statue, he pours the molten metal into a void whose shape is that of the final, solid product. The block containing the void is called the 'mould', and the statue which it produces, the 'cast'.

These metalworking terms have been used to describe fossils by analogy, but since fossils are more complex objects than simple moulds and casts, it has been necessary to refine the meanings slightly.

Let us take as an example an aragonitic shell which becomes fossilized in fine sediment (Fig. 1.2). The soft parts of the animal decay, leaving an internal space which commonly becomes filled with sediment. The sediment now lying around the outside of the shell conforms to the shape and pattern of its surface. Likewise, the sediment filling the body cavity conforms to the shape and pattern of the inner shell surface.

Now, as commonly happens, the shell is dissolved out to leave a void. This, like the mould of the foundryman, is the precise shape of the original fossil. The sediment which bears the imprint of the shell's outer surface is referred to as the **external mould**. The sediment which filled the body cavity and bears the imprint of the internal features, is referred to as the **internal mould**. (It is also sometimes called by the German name **steinkern** which means 'stone kernel'.)

If, subsequently, percolating solutions should precipitate some mineral within the void, then the resulting replica of the fossil is called a **cast**. If the void is left unfilled, it may be possible for the palaeontologist to produce an artificial replica using latex, so as to study all the details of the original.

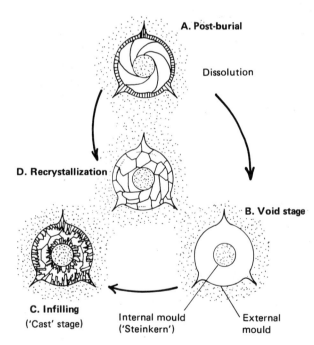

FIGURE 1.2. A is an aragonite fossil seen here in cross-section after burial, with sediment infilling the body chamber and enclosing the external surface. B shows the same fossil after dissolution, which has removed all original shell material and left a void. This void is infilled in C by precipitation of calcite. The replica so produced is a 'cast', made between the internal and external moulds. Alternatively, the aragonitic fabric of A may undergo direct recrystallization without dissolution and the intervention of a void stage. In such cases, some traces of the original crystal fabric may be discerned in the new (D).

1.8 How do fossils 'date' rocks?

Stratigraphy is the study of how layered rocks and

their contained fossils are distributed in space and time. At its heart lies the **Law of Superposition**, which states that in any pile of sedimentary rocks which have not been disturbed by earth movements, the oldest bed will lie at the bottom and the youngest at the top.

But beds of rock are very frequently found to have been disturbed by folds and faults; and over large distances, any particular bed may change in thickness and composition. Thus it may not be possible to match beds of the same age in distant places, and even if we could, the fact that the rocks have the same composition is no guarantee that they were formed at the same *time*.

Fossils provide us with an answer to this problem. Since life is forever changing through time, with species continuously arising and becoming extinct, it is possible to identify strata by means of their enclosed fossils, since these fossils may not occur in beds either older or younger. In practice, strata are actually zoned by the first appearance of the **zone fossil**, an event which extends like a 'time contour' through rocks of that age everywhere.

So the ideal stratigraphic marker fossils will be those with a worldwide distribution, occurring in all sediments irrespective of depositional environment. In the sea, free-living swimmers or floaters will live independent of bottom conditions, and should therefore be good zone fossils if preserved. Wind-borne spores and pollen would be even more widely distributed, being carried across the entire globe in the atmosphere, to fall into freshwater environments as well as the sea.

The second important requirement of zonal index fossils is that species should be short-lived, i.e. the forms should display rapid evolution. This is necessary for the correlation to be a fine one, indicating precise equivalences rather than broad ones.

In the Lower Palaeozoic the graptolites provide perhaps the best example of zonal macrofossils, and in the Mesozoic the ammonites provide a time stratigraphy which is almost unrivalled. The first appearance of a particular species defines the base of a **zone**, which is that thickness of rock between the appearance of one zone fossil and the arrival of the next. This zone is named after its characteristic fossil; thus the Jurassic ammonite *Zigzagiceras zigzag* defines the '*zigzag* zone'. (Remember that the term 'zone' refers to rock thicknesses. The time during which the zone fossil held its sway is known as a **chron**.)

There are all manner of zonations, each based on different fossil groups, and each with its own merits in rocks of different ages, or in different circumstances of work. For example, ammonite zones are of very little use to the oil industry, since the chances of a narrow oil well passing through a specimen of *Zigzagiceras*, and of the fossil then surviving the action of the drill bit, are distinctly remote. But microfossils can be transported to the surface in rock fragments, and their abundance means that more or less any sample of cuttings taken at random will contain them.

So fossils can be used to label all rocks of the same age with the same marks, so that any sedimentary rock (unless it is totally unfossiliferous) can be placed correctly in the stratigraphic column and correlated with its time equivalents.

This progress is called **biostratigraphy**, and it is significant precisely because it is inseparably tied to the real ages of rock bodies. We have said that rock type, on the other hand, is a poor indication of age equivalence. This is because a sediment is the product of certain conditions of physics, chemistry and biology obtaining at a particular place. But those conditions can be duplicated at any time, producing rocks which may look the same, but which are of totally different ages. Also, the set of conditions producing a particular deposit in one place may migrate with time.

Think of a sandbank, some distance off a shoreline (Fig. 1.3). If the land is sinking, then the shoreline will move inland as the transgression proceeds, and all the offshore sedimentary conditions will do likewise. So the sediment laid down by the sandbank will be progressively younger in the direction of the transgression, even though it will come to be preserved as a continuous stratum with consistent lithology throughout. Only the examination of zone fossils will tell the geologist that he is dealing with a 'diachronous' deposit.

Lastly, it should be noted that the dating of rocks by contained fossils is relative, not absolute. The method tells the geologist which unit is the younger, which the older, but not how old, or how much younger. That we know the ages in years of the various geological periods (Table 1.1) is a recent innovation brought in by the technique of radiometric dating. Its accuracy, however, is very poor indeed, and it is never used in the correlation of sediments.

1.9 What are fossils 'occurring *in situ*'?

The Latin term *in situ* simply means 'in place'. Therefore, organisms which are fossilized in the place where they lived and died are said to occur *in situ*. Fossils which have been moved from their habitat to another area for burial are said to be 'transported'. For example, tree trunks may be washed down-river into the sea, or shallow-water bivalves may be exhumed by storms and swept into deeper water for burial.

Strictly, then, the term *in situ* can only be applied to organisms which live in particular places. Free-swimming or free-floating organisms such as graptolites or ammonites therefore cannot possibly be preserved *in situ*, since to be preserved at all they must necessarily be removed from their usual habitat (the water column).

Aquatic organisms which burrow in, or crawl about on, the sediment are called **benthos**. Benthic organ-

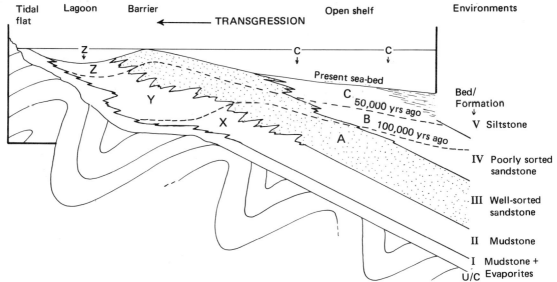

FIGURE 1.3. A marine transgression is in progress, leading to the drowning of the land and the migration of belts of sedimentation across the old eroded land surface (plane of unconformity). Each belt lays down a continuous sheet of sediment which will appear in the stratigraphic succession as a bed or formation with consistent lithological characteristics (I) to (V). However, each bed or formation will not be of the same age everywhere, but will young in the direction of transgression (onlap). Examine the heavy dashed lines. These are 'time lines' connecting parts of the sequence of similar age. They mirror the form of the present sea bed. In the open-shelf sediments, zone index fossils A, B and C delimit the bases of their respective zones and from their distribution enable the stratigrapher to work out the relative ages of sediments where lithological character is unreliable (as it nearly always is). Within the lagoon, the zone fossils do not occur (perhaps because the water there is too saline). But fortunately accessory fossils X, Y and Z can be used, since their equivalence may be established across the barrier sands which interfinger with the lagoon sediments. The mud-flat sediments (Formation I), however, contain no useful index fossils, and so their relative age is rather more difficult to assess directly.

isms often display localized distribution, according to the conditions prevailing on the sea bed. Organisms which float passively, as the graptolites are widely believed to have done, are **plankton**. Forms which, like the fishes, swim actively from place to place are known as **nektic**.

1.10 What are 'derived fossils'?

The erosion of a fossiliferous rock may release fossils into streams or, via the beaches, into the sea. These will be incorporated in new sediments, alongside the skeletal remains of the animals living there, and so could be a source of confusion when a stratigrapher tries to date the younger sediments. Happily, this is not a common problem, but fossils recycled in this way are called 'derived' fossils.

1.11 How may fossils be used as indicators of environment?

Certain animals are known to have very definite environmental requirements and are therefore very selective about where they live. Modern corals, for example, only survive above 90 metres depth in the open ocean, and they only thrive and build reefs above about 40 metres. Moreover, the water must not drop below about 20°C, and they can tolerate neither cloudy water nor any deviation below normal salinity.

By the principle of uniformity, wherever we find fossil reefs composed of modern (scleractinian) corals, we can say that conditions like these also prevailed in the ancient. Unfortunately, the scleractinian corals only date from the Mesozoic, so when we consider the Palaeozoic coral groups (which may have had different requirements) we cannot be so confident in our extrapolations.

Fossils may be used to show how environmental conditions in an area have changed. In the sediments of the Thames dating from the period before the final glaciation of Britain, certain freshwater molluscs are found which live today in warm, calcareous, fast-flowing rivers in southern Europe. This tells us that conditions in England at that time were considerably different from those of today.

Recently, it has been discovered that the ratio of the two oxygen isotopes O^{16} and O^{18} found in the shells of marine invertebrates reflects the temperature of the water in which they live. Warmer conditions encourage the increase of the heavier, O^{18} isotope relative to its lighter counterpart. Thus, by the use of special equipment to detect this ratio in fossil shell material, we may learn the temperatures which prevailed in ancient oceans.

Corals grow a little every day. The daily growth bands are grouped into monthly bands (reflecting the time taken for a new moon to wax full and then wane again to its minimum). It is known that the effects of tidal movement in the earth's oceans contribute to a

TABLE 1.1. *Stratigraphic Reference Chart, with approximate dates in millions of years before present (vertical scale not proportional with duration)*

Phanerozoic	Caenozoic	Quaternary	Holocene (Recent)		
			Pleistocene		2
		Tertiary	Neogene	Pliocene	7
				Miocene	26
			Palaeogene	Oligocene	38
				Eocene	63
				Palaeocene	65
	Mesozoic	Cretaceous	Upper Cretaceous	Maastrichtian	
				Senonian	
				Turonian	
				Cenomanian	
			Lower Cretaceous	Albian	
				Aptian	
				Barremian	
				Hauterivian	
				Valanginian	
				Berriasian (Ryazanian)	135
		Jurassic	Upper Jurassic	Portlandian	
				Kimmeridgian	
				Oxfordian	
				Callovian	
			Middle Jurassic	Bathonian	
				Bajocian	
				Aalenian	
			Lower Jurassic	Toarcian	
				Pliensbachian	
				Sinemurian	
				Hettangian	190
		Triassic	Rhaetic		
			Keuper		
			Muschelkalk		
			Bunter		235
	Palaeozoic	Permian	Zechstein	Tartarian	
				Kazanian	
			Upper Rothliegendes	Kungurian	
			Lower Rothliegendes	Artinskian	
				Sakmarian	280
		Carboniferous	Upper Carboniferous (Silesian)	Stephanian	
				Westphalian	
				Namurian	
			Lower Carboniferous (Dinantian)	Visean	
				Tournaisian	345
		Devonian	Upper Devonian		
			Middle Devonian		
			Lower Devonian		395
		Silurian	Pridoli (Downtonian)		
			Ludlow		
			Wenlock		
			Llandovery		430
		Ordovician	Bala (Ashgill/Caradoc)		
			Llandeilo		
			Llanvirn		
			Arenig		
			Tremadoc		500
		Cambrian	Upper Cambrian		
			Middle Cambrian		
			Lower Cambrian		570
Proterozoic		Precambrian			?3000
Azoic					?4600

slowing down of the planet's rotation through time. Therefore, the number of days in the year (which is the period over which the earth orbits once around the sun) has been decreasing steadily through geological time. Physicists calculate that there were probably 399 days in the Devonian year, and by dividing this figure by 30.6 (which is the average number of daily growth increments in each 'monthly' band of Devonian corals) we find that those 399 short days were grouped into about 13 months. So, fossils can also tell us about the solar system!

1.12 What is taxonomy?

The first task of science is to group observed phenomena and described objects into meaningful categories, so that their unifying characteristics may be made clear and even used as a basis for prediction. It is the expression of a need which mankind has always had, which is to see order and meaning in the diversity of things around him.

And the diversity of living things is huge. Well over a million species of animals alone have already been described, and some system of classification is necessary, even if only for ease of reference. How do we start to put animals and plants in meaningful order?

Classifications are man-made things, and they may be erected on any basis we care to choose. We could classify motor cars, for example, on their colour, their engine capacity, or whether or not they have sun roofs; it all depends on what you want your classification to do. Most of us, when asked about our cars, mention first of all the maker's name. This is a form of **generic** classification, which means that it is based upon origin and common ancestry. It is a natural choice for living things, since our own informal classification of people — the names we give ourselves — are generic. All of us belong to a greater group of inter-related beings, a group denoted by our surnames.

The fundamental unit of biological classification is the **species**. When we recognize dogs, cats, blackbirds, gerbils and hamsters, we are instinctively recognizing different species. Apart from minor differences, all members of a species share a range of features not shared by any other species. In other words, they consistently resemble each other more than members of any other group.

Inherent in this unity of appearance and shape is the notion that all members of a species may interbreed successfully to produce viable offspring in the natural state. The minor differences which we can see in the uniqueness of each individual human being are no barrier to successful mating. However, a human being could not mate successfully with any other animal, not even one of the apes to which man is most closely related.

Sometimes in captivity animals of different but closely related species may interbreed and produce offspring which, to some extent, share the properties of their parents and which may also have distinct characteristics of their own. The horse and the donkey will produce the mule, and the canary may be mated with the goldfinch to produce the so-called 'canary mule', much prized for its sweet song. These animals are called **hybrids**. They are very rare amongst animals in nature, and do not affect our definition of the species, because they are not viable: that is, they are sterile, and cannot themselves produce offspring.

Once defined, similar species may be grouped together in larger categories whose member species more closely resemble each other (i.e. have more things in common) than species of another group. The horse and the donkey, though obviously not of the same species, are nonetheless very similar and may be deemed to belong to the same larger group. This larger group is called the **genus** (pl. **genera**).

Therefore, when we call an animal by its scientific name, we use a system of two words called a **binomen**. The first name, which is always written with a capital letter, denotes the genus and is called the 'generic name'. The species or 'specific' name comes after, and has no capital letter. In printed books, such names are always written in italics. When you write them yourself, or have your manuscript typed, they must be underlined. The names themselves are chosen to be descriptive, though they may also be dedicatory. Descriptive names are taken from Latin or Greek, and even when a species is named after somebody, the name must be 'Latinized'. Here are some examples.

Canis familiaris	The domestic dog. *Canis* = dog (L); *familiaris* = 'belonging or relating to a family or household' (L).
Homo sapiens	The human being. *Homo* = a man (L); *sapiens* = 'sensible, judicious, wise' (L).
Protospongia fenestrata	A sponge from the Cambrian. *Protos* = 'the first' (Gr); *spongia* = sponge (L); *fenestrata* = 'with openings': (*fenestra* = window (L)).
Stenoscisma humbletonensis	A brachiopod of the Permian. 'Steno' = narrow: (*stenos*, Gr) 'scisma' = split or cleft: (*schisma schismatos*, Gr); humbletonensis = 'from Humbleton': '-ensis' is a suffix used to mean 'from' a particular place. 'The brachiopod with the narrow cleft, from Humbleton'.
Arachnophyllum murchisoni	A coral from the Silurian. 'Arachno' refers to a spider's web. In Greek legend, Arachne was a Lydian maiden who entered a spinning contest with the god-

dess Minerva and was turned into a spider. The scientific name for spiders is 'Arachnida'. 'phyllum' means 'leaf-like': (*phyllon* = leaf (Gr)); *murchisoni* = 'of Murchison', after Sir Roderick Impey Murchison (1792–1871), founder of the Silurian and Permian Systems and onetime Director General of the Geological Survey of Great Britain.

This system of naming has been with us since it was set out by the Swedish botanist Carl von Linné (1707–78), more commonly known by his Latinized name Linnaeus, in his book *Systema Naturae* (The System of Nature) in 1758. But it does not end at the genus level. Similar genera are grouped into families, families into orders, orders into classes and classes into the greatest division of all, the Phylum. (The names of these groups are slightly different in botany, but the principle of the classification is the same.)

There are over twenty phyla in the Kingdom Animalia, some containing millions of species, some only a few. And alongside the animal and plant kingdoms (familiar to all of us) there are three other kingdoms comprising organisms which cannot be described as either animals or plants; these are discussed in Chapter 10.

There has always been some degree of unease about the fact that some of the phyla have seemed ridiculously big, and this has led recently to the creation of a 'superphylum' to enclose all invertebrate animals with jointed appendages. So the new Superphylum Arthropoda (it used to be called simply 'phylum') now contains three-quarters of a million insects (Phylum Insecta, formerly a class of the old phylum) as well as the king crabs, spiders, scorpions, centipedes, millipedes, true crabs, lobsters, shrimps, water fleas, ostracods, copepods and barnacles — together with many more minor groups and several major extinct groups such as the trilobites. In contrast, the Phylum Priapulida contains a mere eight species.

To sum up, we have a basic unit of taxonomy called the species which has a recognizable existence in the real world. These species are grouped in progressively higher categories from genera to phyla, but these higher groups are frequently changed as the state of our knowledge changes. This means that species may move from genus to genus, phyla be broken up into new phyla, or degraded to class level, or elevated to the rank of superphylum. Higher classification is an *interpretation*. Here is the classification of our own species as it stands at present.

Kingdom Animalia		all animals
(Subkingdom Metazoa)		all animals consisting of more than one cell
Phylum Chordata		possessing an axial, dorsal nerve cord
(Subphylum Vertebrata)		nerve cord surrounded by a bony spine
Class Mammalia		true hair, nursing of young, brain of advanced type
Order Primates		mostly tree-dwelling placental mammals
(Suborder Anthropoidea)		monkeys, apes and man
Family Hominidae		man and his immediate ancestors
Genus *Homo*		man
Species *sapiens*		sensible, judicious, wise

So the species is a concept created by biologists: but a biologist deals only with life as it is now. A biologist can see species alive in nature, and can watch them interbreeding. Palaeontologists, on the other hand, can not. Moreover, we have to deal with species as they progress through time. We have to think of species in four dimensions, not just in the conventional three.

The biologist's species is the **biospecies**, the true and full expression of Linnaeus's concept. But fossil species are described by form alone, and are defined solely on their morphology. They are therefore termed **morphospecies**. And since fossil species also have a distribution in the time dimension, from their inception to their eventual extinction, they are also **chronospecies**.

So palaeontologists can never be sure that their morphospecies were true biospecies — though it is a reasonable assumption, particularly since most living species are clearly recognizable on form. But remember that not only must a palaeontologist set limits on the morphological range of a species at any one time; he must also set limits upon the evolutionary variation in form seen *through* time. Progressive changes in form can therefore become so pronounced that the palaeontologist may deem the form to have become sufficiently different for it to receive a new specific name.

1.13 Who creates the classification?

When a new species of fossil is discovered, a new name must be chosen. We have seen how it can be descriptive, or refer to the place where it was found, or be named after someone (this can be anyone — even the author's pet dog — but it cannot be the author himself!).

The description of the new species, together with abundant diagrams and photographs, is released in a special scientific journal such as *Palaeontology*, which is published by the Palaeontological Association. The description, in the form of a scientific paper, is examined by the editors of the journal and by a number of referees who are chosen for their special knowledge of the particular group to which the new species belongs.

Once the manuscript has been corrected and the revised version accepted, it will (in due course) appear in print. This is how the basic work of palaeontology is carried out. Without it, and without adequate classification, no science would be possible.

In order that future scientists may re-examine the material, all the fossils illustrated in the paper are deposited in a museum. In most cases this selection will include one special specimen chosen by the author to represent the characteristics of the new species to the fullest degree. This is the 'type specimen'. Its location and its museum number will appear in the paper when it is published.

Because taxonomy is so important, no taxonomic work is recognized unless these procedures are carried out. There is a complex set of rules governing the establishment of classification and ensuring that it remains standard. It is known as the **International Code of Zoological Nomenclature** (there is a separate one for botanists) laid down by an international commission.

1.14 What is 'numerical taxonomy'?

When biologists first began classifying organisms, they found that they were able to perform this task by referring particularly to certain features which seemed to offer the closest approximation to the true derivation of the animal or plant. Linnaeus, for example, used floral attributes to classify flowering plants, rather than, say, the shape of their leaves. Ammonite specialists use the complex pattern of suturing which marks the junction between the outer shell and its internal walls. These things are thought by specialists to be less sensitive to short-term changes, and therefore more expressive of an organism's origins.

Recently, there has been a move away from this style of approach, and the new methods are finding converts in palaeontology also. Critics of the classical methods see the choice of special (or 'weighted') characters as subjective and unscientific. Rather than making 'before-the-fact' decisions about which features are important, they prefer to amass every observable measurement and characteristic about the specimens under review, and then to use a computer to make an objective assessment of how similar one set of characters is to another, without 'weighting' any particular feature beforehand. The *Oxford English Dictionary* defines 'objectivity' as 'treating a subject so as to exhibit the actual facts, not coloured by the feelings or opinions of the writer'. This is what numerical taxonomy aims to do.

Although it is a new field, the concepts which it uses originated with a French botanist called Michel Adanson (1727–1806). His ideas, however, could not be realized practically until computers had been developed that could deal with the immense amount of laborious calculation. Using an appropriate program, a computer can analyse similarity over the whole range of attributes, plotting its results in a way which displays these resemblances graphically (Fig. 1.4). There are many types of array to choose from, some, like the dendrogram shown, using a 'family tree' approach. Others may come in the form of an array of points, whose closeness of cluster represents degree of similarity.

So, in a typical array, the most closely spaced points may represent one species. Each species cluster may occur in one of two groups of clusters, and these groupings may represent genera. The numerical method does not tell the taxonomist where to draw the line between species, or between species and genera. This is left to subjective judgement, because ultimately all taxonomy must be a matter for fallible human decision. Numerical methods merely restrict

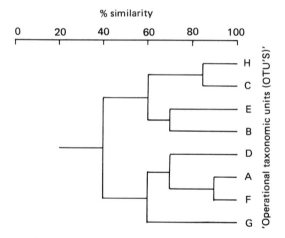

FIGURE 1.4. This diagram is known as a 'dendrogram', and it is widely used in numerical taxonomy. It expresses the similarity at different percentage levels between an array of specimens A–H. After the computer has determined the percentage correspondence between the measured characters on each specimen with those on every other, it expresses this in terms of the branching pattern. Thus, H and C are 85% similar, E and B 70%, while [H + C] and [E + B] are collectively 60% alike. The whole array has separated into two distinct groups which overall are only 40% similar. The taxonomist must decide where to draw the specific and generic boundaries. Are [H + C], [E + B], D, [A + F] and G each individual species falling into two genera, or are [HCEB] and [DAFG] two separate species in one genus? It is here that the final and inevitable element of subjectivity enters.

the degree to which such subjective assessments affect the outcome.

One of the major advantages of the method is that, by removing the necessity for years of personal acquaintance with a particular group, the taxonomy can be worked out much more quickly. The work is also 'repeatable' — i.e. another scientist can easily check the results and come to the same conclusions with the same data. But it must be said that the new science of taxometrics usually finds itself confirming the work of classical taxonomy. Its advantages are chiefly those of speed.

1.15 Welcome to palaeontology

We hope that the brief outline set out in this chapter has encouraged you to find out more about the science of palaeontology. The next chapter deals with the ways in which life originated on earth, and describes some of the oldest known fossils, from rocks of the Precambrian.

From Chapter 3 on, all the major fossil groups of the Phanerozoic (Cambrian to the present) are described, and Chapter 12 attempts to bring together the major themes which have emerged within a discussion of the principle of evolution.

We hope that you will make use of the many annotated diagrams to make your own scaled drawings of fossil specimens. The easiest way to learn is to see things for yourself, and to record them.

It is also vital to understand the reasons behind the appearances. As a matter of necessity we ask the question 'what?': but only so that we may then ask 'why?'

2.
The Origin of Life and the Earliest Fossils

2.1 When and how might life have begun?

The earth was formed about 4600 million years ago, but it probably spent about 1000 million lifeless years before the first primitive bacterial organisms developed upon it. During that immense period (almost twice as long as the period from the Cambrian to the present) the crust was likely to have been extremely mobile and subject to such cosmic influences as meteor bombardment. There is also evidence that, sometime during this early phase, the entire atmosphere was stripped away and slowly restored as more gases emanated from the planet's interior.

This atmosphere would have been rich in carbon dioxide, carbon monoxide, methane, ammonia, hydrogen and nitrogen. These noxious volcanic fumes stormed across the bare rock and its waste, unable to protect the surface of the earth from the lethal ultraviolet radiation of the sun. How could life have arisen naturally under such conditions? Strangely, the action of lightning discharges (and other sources of energy) upon the primordial atmosphere would have been able to create complex molecules consisting of long carbon atom chains. And these are the compounds upon which the chemistry of life depends.

In a famous experiment, J. J. Miller of the University of Chicago placed a mixture of gases similar to that envisaged for the early atmosphere into a closed vessel together with some water (Fig. 2.1). He then passed a strong electric discharge through the contents of the flask to simulate lightning, and found that after a week, significant amounts of amino acids — the raw material of proteins — had been formed.

Such molecules are unstable today if exposed to the elements, because the atmosphere now contains oxygen. At that time, however, the earth was devoid of free oxygen, and complex organic molecules could survive once they had been synthesized. Miller and others repeated the experiment many times, elaborating it by the addition of other materials which would have been abundant at the time. In this way they succeeded in producing nearly all the amino acids used today by living things. Most interesting of all,

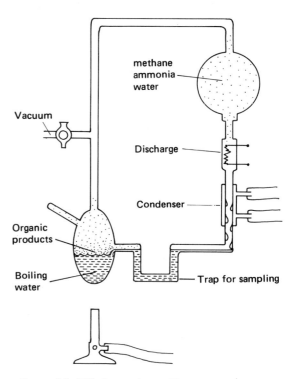

FIGURE 2.1. Miller's experiment. The action of the heavy electrical discharge was to combine the raw materials into various organic compounds, notably amino acids, as well as urea and formic acid.

they found that they never produced any of those amino acids which, although possible to make in the test-tube, are nevertheless never found in nature.

The next step was obviously to see whether these amino acids could be induced to form proteins. Sydney Fox, at the University of Miami, reasoned that such combinations would only be likely if the dilute solutions could be concentrated. Hot, dry conditions were needed — conditions such as might well have occurred around volcanoes. The volcano soon emerged in the minds of many people working on the problem of biogenesis as a likely crucible for the formation of the first lifelike objects, particularly when it was realized that volcanoes are places where torrential rain and electric discharges are encouraged by eruptions.

Fox tried baking some of Miller's organic broth on blocks of Hawaiian lava. He found that the amino acids did indeed join up, forming what he termed 'proteinoids'. But more exciting than this (and quite unexpected) Fox noticed that if the proteinoids were quenched with water, they formed themselves into cell-like globules with double membranes. If water of the right acidity were used, these could even be persuaded to bud and divide just like true cells.

In the Soviet Union, Alexander Oparin created blobs similar to Fox's. He called them 'coacervates', and they were formed by mixing proteins and polypeptides (e.g. albumen and gum arabic). They were especially interesting in that they could selectively absorb substances from the environment. In other words, they were on the road towards being capable of regulating their internal chemistry — a principle of life called **homeostasis**. Oparin introduced catalysts into his coacervates which enabled them to perform simple biochemical reactions such as transforming glucose into starch.

Of course, Fox's microspheres and Oparin's coacervates lacked the blueprinting mechanism which controls the activities of life and assures its continuation — DNA. But experiments since the late fifties and early sixties have gone a long way toward proving that nucleic acids could also have been formed in the same haphazard fashion.

Given the limitless time which was available, it need no longer seem so incredible that natural processes could have created the complex raw materials needed for the activity of biochemical processes within the early cell-like capsules which may have been the precursors of all life on earth.

2.2 What can geology tell us about this early life?

Once the earliest forms of life had developed, their only means of growth and survival would have been to take in the organic molecules around them as 'food'. This could have continued for a little time, but sooner or later the supply would have run out. In the modern world, animals (which ingest other life forms) do not exhaust their resources because ultimately they all rely upon plants, which can make organic matter from inorganic materials using the energy of the sun. For life to develop further from its earliest condition, **photosynthesis** (which is the plants' biochemical 'trick') would have to evolve. Its eventual appearance was one of the most momentous events not only for life on earth, but for the earth itself.

The chemical reactions of photosynthesis release oxygen as a by-product. The plant uses some of it for its own respiration, but production exceeds consumption, and the residue is dumped in the atmosphere. All the oxygen in the modern atmosphere has been put there by plants. It is a highly reactive substance, and when it was first produced in quantity by early plants it would have combined immediately with other substances such as iron, which oxidize readily.

The sudden oxidation event which marked the appearance of plants is recorded in the sediments of the Precambrian. Huge thicknesses of red ironstones tell of massive oxidation taking place worldwide about 3.2 billion years ago. So we can tentatively place the origin of photosynthesizing plants at about that date. Before that, only simple bacteria which could live without oxygen existed, and their fossil remains — about 3.5 billion years old — have been tentatively identified in the rocks of Western Australia.

The appearance of oxygen had one more far-reaching environmental effect. Under the influence of solar ultraviolet, the O_2 molecules were transformed into the unstable molecule of ozone (O_3). This created a layer of ozone in the outer atmosphere which screened off the larger part of incident solar ultraviolet, making the earth far more hospitable to life.

So — the stage was set; the harmful radiation was shut out; plants were building an atmosphere rich in oxygen. Conditions were right for the arrival of a type of organism which fed upon this renewable organic matter, using the oxygen in the air to convert the food into energy — the animals.

2.3 Precambrian animals

The record of animal life in the Precambrian is very scanty indeed. However, rare fossils in certain exceptional localities have enabled us to piece together some sort of picture of these strange creatures.

In the 1940s, in interbedded sands and silts, some delicate impressions were discovered. The site was in the Ediacara Hills of South Australia, and the rocks — which lay demonstrably below the basal Cambrian — were found by radiometric dating to be about 700 million years old.

A thousand or so specimens have now been grouped into roughly thirty species comprising soft corals, jellyfish, worms and others whose affinities are not resolved. Some of these creatures are shown in Fig. 2.2. Their mode of preservation suggests that these animals were stranded upon a muddy shore by retreating tides, to be buried by subsequent sandstone deposition. They serve to show us that life had reached a moderate level of complexity before the beginning of the Cambrian, and the evidence of trace fossils suggests that certain worm-like organisms were active as long ago as 1 billion years. But it remains true that no fossils, even these burrows, are at all common before the end of the Precambrian.

Australia does not have the monopoly on these very early fossils. It was of some interest, for example, to discover that *Rangea* was very similar indeed to the

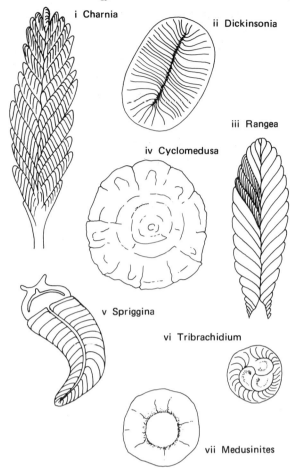

FIGURE 2.2. Some fossils of the Precambrian (not to scale).

the Lower Cambrian has revealed that the introduction of the different fossil groups was not achieved all at the same moment.

But animal life was radiating for the first time into a world where there had been nothing save algae and bacteria for æons. The living world was being established, and no present-day analogues can be found to help us understand what that must have been like — except, perhaps, the growth of bacteria upon a Petri dish of nutrient agar.

Each bacterial cell divides once every few minutes. Growth is slow at first; after the first period, one cell becomes two. Then, two become four, and so on, doubling every time. In no time at all there are millions of cells, yet the individual bacteria are not dividing any faster. Try it on your calculator, multiplying one by two, then three, then four . . . and see how long it takes you to reach a million.

This is the way populations grow when there are no limits to growth. Of course, the rate slows eventually, because of competition for space or food. But perhaps those incredible 10 million years at the beginning of the Phanerozoic were the 'log phase' (Fig. 2.3) of life's development, where instead of individuals, it was new species which were being created.

fossil known as *Charnia*, from rocks of similar age in Charnwood Forest, Leicestershire. Also, fossil jellyfish closely resembling *Cyclomedusa* and *Medusinites* have recently been unearthed from rocks south of Carmarthen, Dyfed, together with another form not unlike *Tribrachidium*.

So, animal life did not begin suddenly, from nothing, at the base of the Cambrian. Nevertheless, it is very remarkable that before the start of that period, there had been thousands of millions of years during which nothing more complex than algae and bacteria had existed; yet in a mere 10 million years from the opening of the Phanerozoic (see below) nearly all the basic animal types now seen on earth had already come into being. The groundplan of all subsequent animal development, it seems, was set in a geological instant.

2.4 The Phanerozoic begins

The term 'phanerozoic' means 'evident life' and refers to the period from the beginning of the Cambrian to the present (Table 1.1). Its beginning certainly was sudden from a geological point of view, but not as sudden as it once seemed. For in addition to the Precambrian faunas, more detailed stratigraphy of

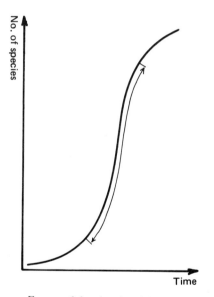

FIGURE 2.3. A sigmoid curve illustrating the 'log phase' — a period where logarithmic increase in numbers takes place. Such an effect may help to explain the sudden development of life at the base of the Cambrian.

But there are other factors to consider. For a long time, the sudden appearance of fossils has been attributed to the first development of hard parts. In other words, we are witnessing a sudden rise in preservation potential rather than in life itself.

The argument is partly valid. Certainly, hard skeletons were not found in the Ediacara Fauna, nor in any other from that time. Hard exteriors may have arisen 'by accident' in the first instance, perhaps with the disposal of certain waste products in the skin. But once one group began to exploit the advantages of a hard shell, competition would ensure that similar devices were soon developed in others. For what are the advantages? Protection, certainly: but a skeleton also improves general efficiency — particularly of movement, since it allows muscles to act against solid anchorages. These and other biological advantages would strongly favour 'skeletonized' groups.

But hitherto, the assumption of this argument has been that, prior to hard parts, the world was populated with soft-bodied equivalents. This is probably not true. Firstly, no such 'equivalents' have turned up; secondly, it is not possible to separate in this way the 'hard parts' from the kinds of animals which use them. Without their hard parts, the biology of skeletonized animals would not make sense. You cannot have a soft-bodied arthropod. Such animals need their hard parts to be the way they are, and without them their life would be impossible.

To sum up, then, the sudden appearance of fossils at the opening of the Phanerozoic is partly an illusion, because of the increased preservation potential. But the origin of hard skeletons also amounted to such an innovation in animal structure that it led to the explosive radiation of new and rapidly developing groups into a world where there were no constraints of space or available food. Revolutionary progress in design and unlimited opportunities for expansion made the Lower Cambrian the single most important period in the history of life.

3.
Trilobites

Superphylum Arthropoda Phylum Trilobita (L. Camb.–U. Perm.)
[*Note*: This classification has been put forward by Manton (1973, 1977) and is different from that in the *Treatise*, which regards the Arthropoda as a Phylum of which the Trilobita forms one Class.]

3.1 Introduction

The Superphylum Arthropoda is an enormous group of invertebrate animals all characterized by paired, jointed limbs and a hard external skeleton (exoskeleton) which must be moulted periodically to allow for growth. Arthropods have well-developed internal systems (nervous, digestive, circulatory, etc.) and are highly successful. At the present time, 75% of all known animals belong to this superphylum, and 75% of these are included in one phylum, the Insecta. The success of insects dates from the Cretaceous, when flowering plants first appeared in numbers. But in the Lower Palaeozoic it was the trilobites which were the chief representatives of arthropod design.

The proliferation of the trilobites was the first massive radiation of arthropods into the marine realm, where they lived mostly in fairly shallow waters crawling and grubbing along the bottom. The great majority lived this kind of benthic existence, although a few are known to have been free-swimming, or nektic. And although most trilobites were less than 10 cm long, some have been discovered which were almost 70 cm from end to end. They are without doubt the most important fossil arthropods, but there are many others (see Table 3.1).

TABLE 3.1. *Fossil Arthropods (excluding Trilobites)*

Group	Geological range	Examples
Crustacea	L. Cambrian–Recent	Ostracods (see Ch. 10), shrimps, crabs, barnacles
Merostomata	L. Cambrian–Recent	King Crabs, eurypterids (fossils only)
Arachnida	Silurian–Recent	Spiders, scorpions
Myriapoda	Devonian–Recent	Centipedes, millipedes
Insecta	Devonian–Recent	All insects: silverfish, springtails, mayflies, dragonflies, butterflies, caddis flies, stoneflies, houseflies, grasshoppers, locusts, termites, earwigs, aphids, beetles, moths, ants, bees, wasps, fleas, lice and thrips

is reached, while others go on growing and ecdyse more or less continuously. Trilobites apparently fell into this second group. This means that for every one trilobite which lived, there were many potentially fossilizable moults produced in addition to the one, eventual, entire body fossil.

3.2 Ecdysis

This is the scientific term for the moulting of the exoskeleton. Because this suit of armour does not expand, it must be sloughed off, and a new, larger one developed. Increase in size is therefore restricted to the 'soft' stage, when the animal is extremely vulnerable to predators. Arthropods may go through this process many times in their lives, and each completed cycle from soft shell to eventual ecdysis is called an **instar**.

Different groups of arthropods have different systems of ecdysis. Some moult only until the adult form

3.3 Morphology

From the name 'trilobite' you will gather that the animal had a three-lobed construction. Figure 3.1 shows that this is the result of two deep longitudinal furrows which divide the dorsal surface into an **axis** and two lateral areas called **pleura** (s. pleuron).

At this point it might be as well to point out that there are precise scientific terms for 'front', 'back', 'top' and 'bottom' when referring to animals and their fossil remains. The animal's back is known as its **dorsal** surface; its underside or belly is referred to as **ventral**. The head region or front end is termed **anterior**, and the rear or tail-end, **posterior**.

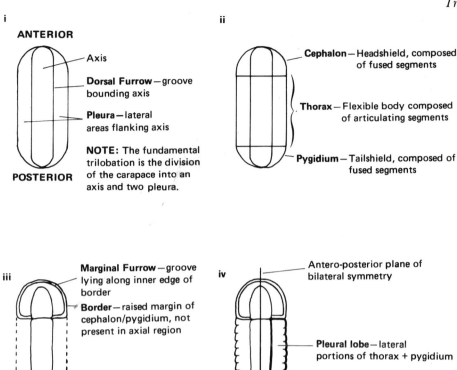

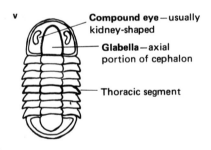

FIGURE 3.1. Basic trilobite morphology. Diagrammatic dorsal views, not to scale.

One of the features common to many lower invertebrates is the **segmentation** of the body. This is a system of construction whereby the body is made up of a sequence of compartments. This pattern is masked in the more highly developed invertebrates, and is perhaps seen to best advantage in worms. Traces of such segmentation are seen in trilobites, which probably evolved from a worm-like ancestor some time in the Precambrian.

Trilobites have a distinct headshield or **cephalon** and a distinct tailshield or **pygidium**, and in both these regions the original segmentation has been suppressed — more completely in the cephalon than in the pygidium — and the exoskeleton has become fused into a rigid plate. Between these two shields lies the **thorax**, where the segments are not fused but articulate with one another, conferring flexibility.

Many trilobites, indeed, were able to roll up (Fig. 3.4), bringing the ventral surfaces of the cephalon and pygidium together. This was probably done for protection in emergencies, and there are special structures in some species to allow the enrollment to be 'locked', so affording even greater security. Many trilobites died in the enrolled state, but by no means all trilobites were able to curl up in this way.

Each segment of the thorax bore a pair of walking legs and a pair of feathery appendages which may have been external gills and are often called **gill branches** (Fig. 3.6). They are described more fully below. There were also legs beneath the cephalon and pygidium, and from the anterior of the headshield there projected a pair of sensory **antennae**.

Details of dorsal morphology commonly seen in trilobites are shown and explained in Figs. 3.1–3.5. Some of these, however, demand further description.

3.3.1 Facial sutures

The **facial suture** is important in classification, but it was also important to the living trilobite, since it was the line along which the cephalon split open during ecdysis. Typically, this suture runs along the front of the glabella, around the eyes and then either towards the **anterior** margin, the **posterior margin** or the **genal angle** (Fig. 3.3).

During moulting, the **fixed cheek** remained attached to the glabella, while the **free cheek** fell away. The **compound eyes** were left attached to the soft shell.

3.3.2 Eyes

The compound eyes of the trilobite are the earliest example of efficient vision in the animal kingdom. Their similarity to the eyes of insects and crustaceans is obvious, but it is doubtful whether they represent an ancestral form of these. More likely, compound eyes developed separately in these different groups.

Each eye is divided into many lenses, each of which covers a small portion of the visual field. In modern arthropods, each lens lies at the top of a long cylindrical structure called an **ommatidium**, whose refractive system focuses light onto a sensory organ called a **rhabdom**. The trilobites probably had a

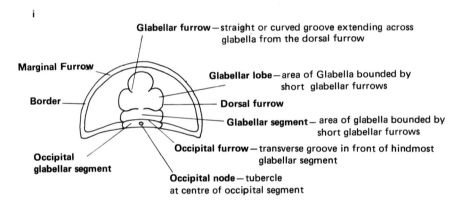
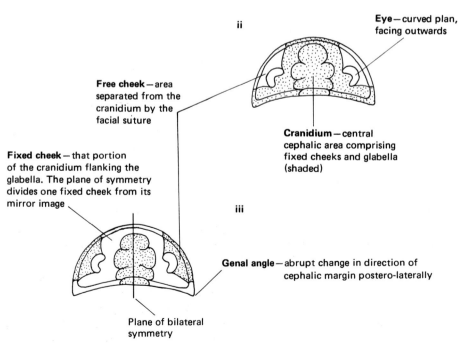

FIGURE 3.2. Cephalic morphology. Diagrammatic dorsal views.

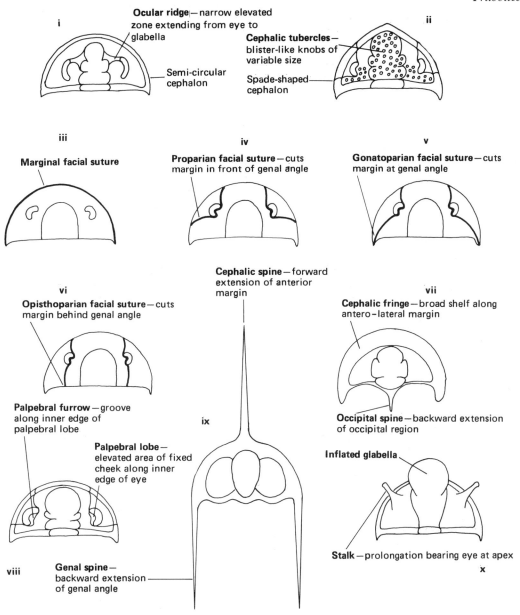

FIGURE 3.3. Cephalic features. Diagrammatic dorsal views (not to scale).

similar system, but as yet little is known of their eyes' internal structure.

The trilobite probably saw the world in a series of vertical strips of light, and the function of its eyes was probably the detection of movement — a function which compound eyes are well suited to fulfil.

Many trilobites had no eyes, while others had eyes so large they even dwarfed the glabella. These differences obviously tell us about mode of life, and are further discussed below. Rare trilobites bore their eyes upon stalks, and some, though apparently possessed of eyes, were probably blind nevertheless. We believe this because the eye structure, when examined in detail, shows variable numbers of lenses from one specimen to another, arranged in a more or less disordered pattern. This is thought to indicate functional degeneracy.

3.3.3 Limbs

Trilobite appendages are only seen in specimens preserved under exceptional conditions. They were **biramous**, that is, composed of two elements, one being a walking leg and the other, lying above it, a gill — though this is not certain, and it may have had other functions too.

Each thoracic segment bore two such biramous couplets, one on each side of the axis. There appears to have been no functional differentiation of limbs as is seen, for example, in crabs whose forelimbs may be

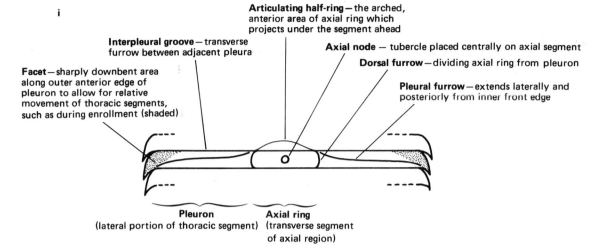

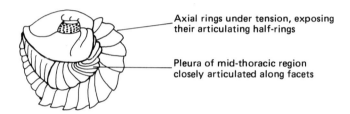

FIGURE 3.4. Thoracic morphology. Diagrammatic dorsal view.

modified into pincers. The only variation in trilobite limbs was the gradual decrease in size towards the posterior, mirroring the taper of the body itself. Even limbs beneath cephalon and pygidium were of the same standard design.

Since trilobite limbs were so uniform, since they appear to have shown little evolutionary development and are, in any case, so rarely preserved, they are not useful either in classification or in determining mode of life. Most of their morphology has been worked out from exceptionally well-preserved fossils in the Burgess Shale (M. Cambrian, British Columbia) and Utica Shale (L. Ordovician, New York) with a few rarities from other localities around the world. The technique of using X-rays to look at fossils within blocks of rock has also shown up many interesting features, not only of the legs but also of the internal struts of the exoskeleton to which muscles were attached.

The two branches of the limb join at an element called the **coxa**. In some trilobites the coxae of a few anterior thoracic segments bear sharp projections which could have had some function in working over the sediment prior to ingestion (Fig. 3.6).

Before proceeding any further with the text, you should familiarize yourself with all the structures shown in the diagrams, and with their scientific names. When you feel confident of your powers of recognition, you should draw your own scaled, annotated diagrams from specimens or replicas. Even copying the diagrams in this book will be of enormous help to you in remembering these forms and their correct terms.

Ultimately, you should aim to be able to draw and label any specimen you are given, after the manner shown in Fig. 3.7. Remember that accurate observation and naming are vital to the identification of species and to the correct understanding of the evolution of fossil structures from one form to another. It is the fundamental job of the palaeontologist, upon which everything else in the science ultimately depends.

3.4 The nature of the trilobite cuticle

The exoskeleton of arthropods is built up of panels of 'cuticle', rather as a motor car is made up of panels of pressed steel. In most arthropods this cuticle is either made of chitin or of chitin strengthened by some mineral.

The trilobite cuticle was rather different, however, being built up of tiny calcite crystals arranged at right angles to the surface and bound together in an organic material. This organic material remains as yet undescribed. It may have been chitin, and many textbooks describe trilobite cuticle as 'mineralized chitin', but no chitin has ever actually been found.

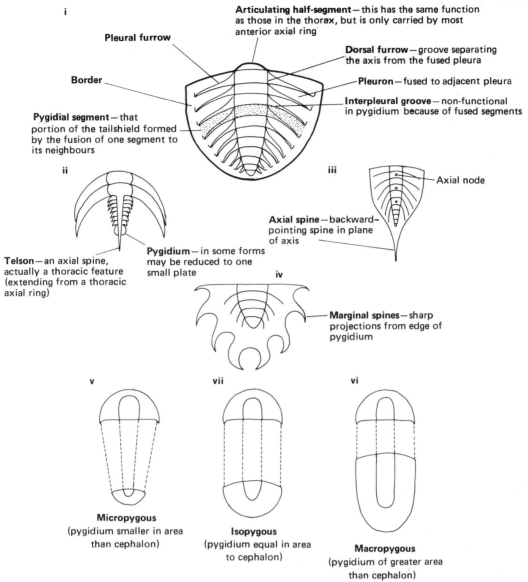

FIGURE 3.5. Pygidial morphology. Diagrammatic dorsal views (not to scale).

3.5 Life cycle and development

Although none has ever been firmly identified, it is presumed that the trilobite began as an egg, from which a larva eventually hatched. The first larval stage which was capable of preservation (and of which, therefore, there is evidence) is known as the **protaspis**. This stage encompasses all growth up to the development of a transverse ridge and furrow which divides the cephalon from the rest of the body. Protaspid larvae were 0.25–1 mm in size.

During the second phase, called the **meraspis**, the posterior portion of the body developed recognizable thoracic segments one by one. The meraspis stage is therefore divisible into **degrees**, beginning with the formation of the ridge and furrow (degree 0) and ending with the number of the last-formed thoracic segment. As each new segment formed, the larva had to undergo ecdysis. Once the final segment was made, the adult form was complete, and this stage of maturity is called the **holaspis**, which means 'complete shield'. Growth continued, however, proceeding by further ecdyses until the death of the individual. The process of development during larval growth is called **ontogeny**.

Many changes took place during this ontogeny. From smallest protaspis to largest holaspis (encompassing therefore ontogenetic and mature growth) we may witness a 400-fold increase in size — although a 50-fold increase is more common. During larval development, the gradual formation of the eyes took place, as did their migration from the cephalic margin to their final sites. In some species, the facial suture began as a proparian type, and migrated during ontogeny through a gonatoparian state to an eventual opisthoparian condition.

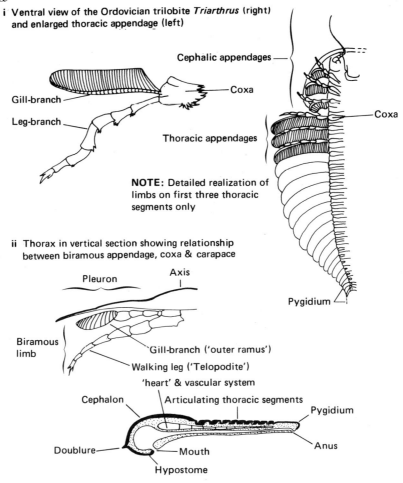

FIGURE 3.6. Trilobite limbs and internal anatomy (not to scale).

Larval stages are useful in taxonomy, because forms with apparently very different holaspids may be very much alike in their early stages. This can often be taken as indicating a common descent.

3.6 Classification

The question of what makes a satisfactory classification was touched on in Chapter 1, and there have been many attempts to create one for the trilobites. For example, one early system used certain heavily weighted characters (notably, the type of facial suture) to create three Orders. It was assumed that the suture types (proparian, opisthoparian, gonatoparian) were natural labels which reflected three distinct phylogenetic lines. The discovery of larval stages whose ontogeny involved the migration of sutures from one condition to another (mentioned above) severely weakened the case for retaining this old scheme.

As is so often the case in palaeontology, we have to try to classify together those forms which, quite simply, look the same. This is a practical thing to do, because it makes identification easy and memorable. Whether these groups based on appearance (**phenetic** groups) actually reflect natural **phylogenetic** or genetically related categories is, and will remain, a matter for debate.

The classification used here is based on that devised by the authors of Volume O (Trilobites) of the *Treatise on Invertebrate Paleontology*, edited by the series' founder, R. C. Moore. The Treatise is published by the University of Kansas, and while its pronouncements do not quite possess the authority of Holy Writ, it is the nearest thing that palaeontologists have to a Bible. Its massive effort aims to cover all fields of the subject and has harnessed the services of specialist authors all over the world: providing what is best described as a state-of-the-art consensus of expert opinion.

The Treatise classification recognises fifteen different types of suture, uses the features of the glabella and also pays attention to the pygidium. But, happily, the Orders are easily recognizable, and Figs. 3.8–3.12 set out their basic features. These are also set out for ease of reference in Table 3.2.

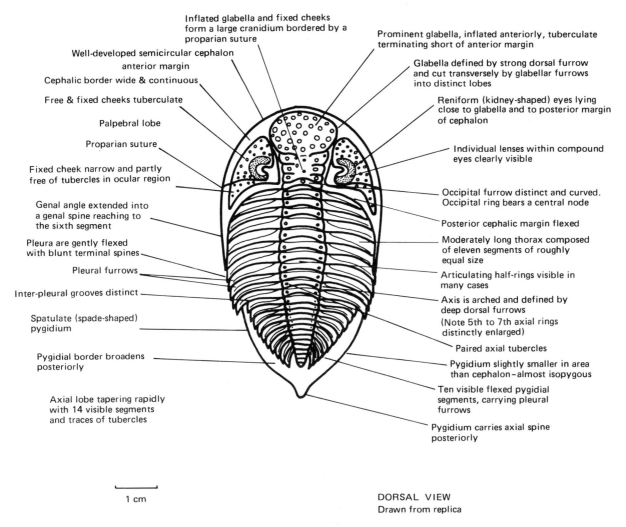

FIGURE 3.7. *Dalmanites myops*.

3.7 Mode of life

The many forms seen in trilobites probably reflect different modes of life, and palaeontologists have long been tempted to make speculative suggestions as to what features correspond to which life style. With increasing knowledge of not only general skeletal structure but also of the microstructure of its surface and surface features, it is now at last becoming possible to support or disprove many of these interpretations which have lain around inconclusively for so long.

Let us look first of all at the eyes. It has been mentioned that the eyes can show various degrees of development or suppression. Trilobites which were blind may have been nocturnal in habit, or may have lived in cavities such as occur in reefs. Alternatively, they might have been burrowers. The absence of eyes is a difficult feature to which to ascribe any definite single cause. It is best weighed with other features before deciding.

For example, those forms with small eyes combined with a wide axial region and smooth carapace, whose thoracic segments appear to have fitted very closely, may have been burrowers. The wide axis possibly reflects the presence of strong muscles, and the smoothness of the exterior aided passage through sediment. There would have been some need for the exclusion of grit from the gaps between the segments, hence their close fit.

Ptychocephalus vigilans (Fig. 3.13 ii) is a trilobite which is often found in life position, its cephalon peeping up above the bedding plane, and the rest of its body buried vertically in the sediment. This was probably a posture adopted while waiting for its prey.

Forms with very large eyes, such as *Cyclopyge*, could not have been benthic, since this life style would have buried their enormous ocular structures in the sediment. Such a degree of all-round vision could only have been of use to a free-swimmer. They may have swum upside down near the surface, so keeping a sharp look-out for prey or predators.

TABLE 3.2. *Trilobite Classification*

Order, Range, No. of Genera	Sub-Order and Range	Major Features	Examples
Agnostida L. Camb–U. Camb 79	Eodiscina L–U. Camb	Very small forms with isopygous form. Three thoracic segments. Mostly blind	*Eodiscus*
	Agnostina L. Camb–U. Ord	As above, only with two thoracic segments. Mostly blind	*Agnostus*
Redlichiida L–M. Camb 107	Olenellina L. Camb	Large semicircular cephalon, strong genal spines and numerous thoracic segments. Pygidium very small, eyes large.	*Olenellus*
	Redlichiina L–M. Camb	As above, with opisthoparian sutures.	*Redlichia Paradoxides*
Corynexochida L–U. Camb 73		This group has a rather poor common identity. The glabella is of varied shape, sutures opisthoparian and with branches sub-parallel. Thorax has 7 or 8 segments	*Olenoides Bathyuriscus*
Ptychopariida L. Cam–U. Perm 798	Illaenina L. Ord–U. Dev	Large and distinctive isopygous or macropygous trilobites, often smooth and inflated in appearance, sutures opisthoparian	*Scutellum, Bumastus Illaenus*
	Asaphina M. Camb–L. Ord	Large, smooth and isopygous, with librigenae (free cheeks) often linked anteriorly. Some forms have exceptionally large, often confluent eyes eg. *Cyclopyge*	*Asaphus, Ogygiocaris*
	Ptychopariina L. Camb–U. Ord	Thorax often the dominant feature. Glabella ends short of anterior margin and pygidium small.	*Olenus, Ptychoparia, Triarthrus*
	Trinucleina L. Ord–M. Sil	Very distinctive, cephalon very large without eyes, long genal spines common, and wide pitted fringe.	*Trinucleus, Onnia, Cryptolithus*
	Harpina U. Camb–U. Dev	Also very distinctive, these have a large headshield like a horseshoe, not regularly pitted, with sturdy genal spines. Eyes small. Many thoracic segments.	*Harpes, Loganopeltis*
Phacopida L. Ord–U. Dev 173	Calymenina L. Ord–M. Dev	Rather homogenous group of gonatoparian forms with lobed glabellae. Thorax has 11–13 segments. Surface features deeply incised.	*Calymene, Trimerus*
	Phacopina L. Ord–U. Dev	Proparian sutures, 11 thoracic segments and well-developed enrollment mechanisms.	*Phacops, Dalmanites, Acaste*
	Cheirurina L. Ord–M. Dev	Variable proparian group of spiky appearance. Pygidial spines common. 8–19 thoracic segments	*Cheirurus, Deiphon*
Lichida L. Ord–U. Dev 173		Very distinctive group of large trilobites. Opisthoparian Pygidium often larger than cephalon having blade or leaf-like pleurae.	*Lichas, Terataspis*
Odontopleurida M. Camb–U. Dev 25		Extremely spinose, opisthoparian. 8–10 thoracic segments, usually with spines. Pygidium very tiny.	*Acidaspis, Leonaspis, Ceratocephala*
Proetida U. Camb–U. Perm		Opisthoparian, with large glabella. Usually isopygous. Sturdy genal spines common, 8–10 thoracic segments	*Proetus, Phillipsia, Bathyurus*

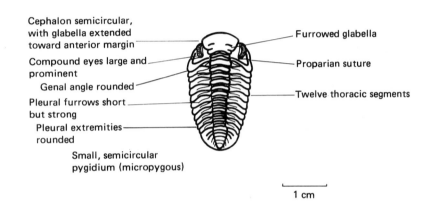

FIGURE 3.8. *Olenoides* and *Ananaspis*.

They are, in fact, often found in deep-water oceanic deposits.

The wide cephalic fringes seen in the trinucleids and harpinids have been a subject for speculation for a long time. The trinucleid brim is made up of two sheets of cuticle joined by roughly cylindrical tubes which look, on the surface, like pits, but which really act as pillars separating the two cuticular sheets. Halfway down these cylinders there is a delicate wall which closes off the tube but for a microscopic central perforation.

What was the function of this brim? Evidence from the traces left by trinucleids resting upon the sediment surface suggests that the animal used its legs to dig a small hole or depression, and that it then positioned itself in the hole with the cephalon angled downwards. Since these traces appear to point consistently in one direction when found together, it is assumed that the animals were aligned towards some prevailing current. Possibly, small sensory hairs (**setae**), situated in the pits along the brim, acted as 'weather vanes' to tell the animal if it was correctly aligned. The implication from this is that trinucleids were filter feeders. The deep anterior arch of the cephalic fringe may have allowed currents to pass under the creature and towards its long, filamentous gill branches.

The brim of the harpids is quite different. It is flat, and its irregularly spaced pits perforate the brim completely. This was probably a resting structure, and may even have acted as a 'snow-shoe' to prevent sinking in soft sediment. Alternatively (or additionally), it could have been a device to render its owner unpalatable to predators.

Spines on trilobites are common, especially in certain orders, and it has been suggested that they acted as baffles to increase water resistance and so help to prevent sinking. In most cases this seems rather unlikely; the spines would probably have had only a marginal effect upon such relatively large creatures.

Perhaps the answer here also lies in unpalatability. Spines on the ventral surface probably acted as supports, but dorsal spines may have to be envisaged in the enrolled position before their protective function can be fully appreciated.

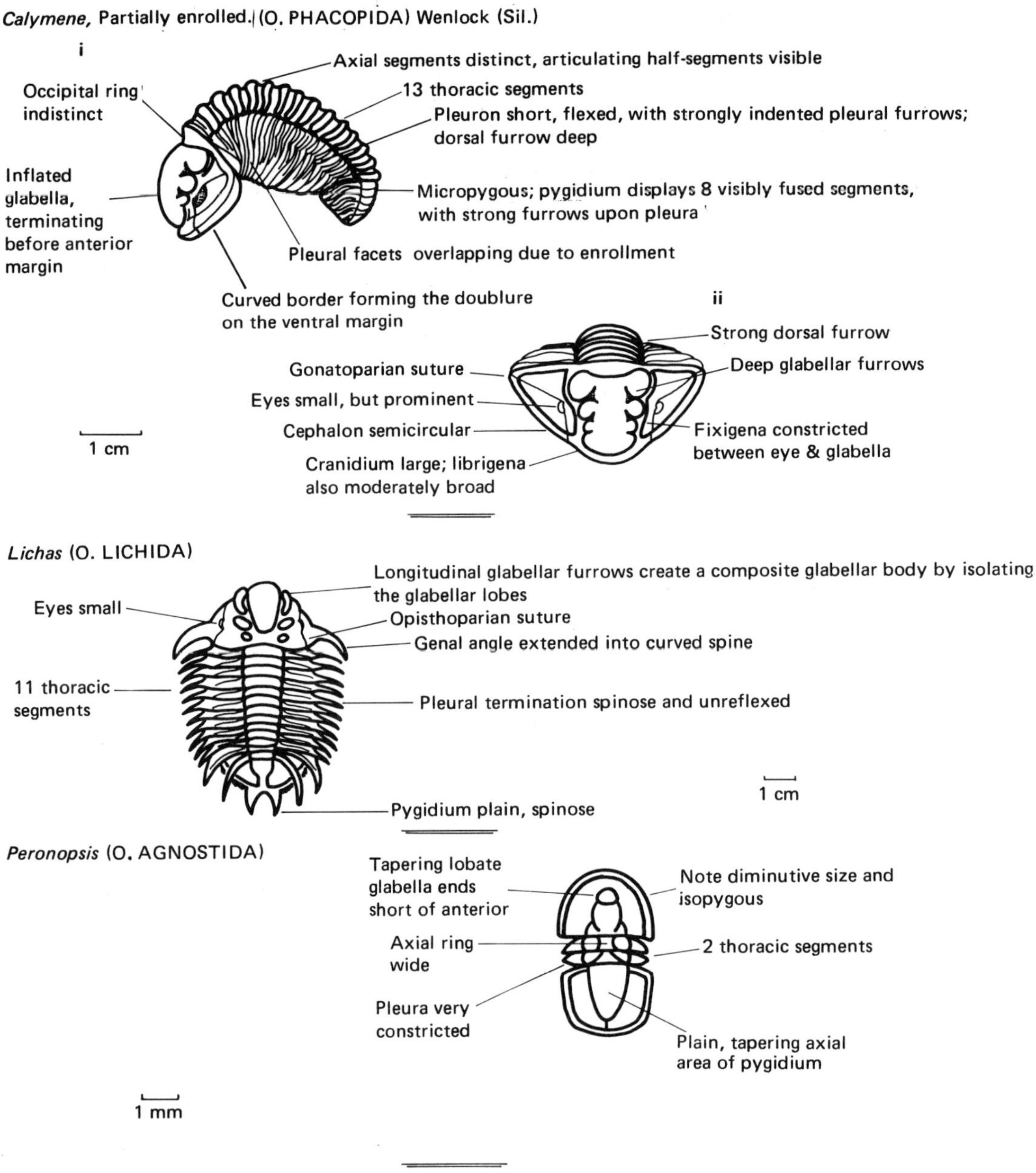

Calymene, Partially enrolled. (O. PHACOPIDA) Wenlock (Sil.)

i

- Occipital ring indistinct
- Inflated glabella, terminating before anterior margin
- Axial segments distinct, articulating half-segments visible
- 13 thoracic segments
- Pleuron short, flexed, with strongly indented pleural furrows; dorsal furrow deep
- Micropygous; pygidium displays 8 visibly fused segments, with strong furrows upon pleura
- Pleural facets overlapping due to enrollment
- Curved border forming the doublure on the ventral margin

1 cm

ii
- Strong dorsal furrow
- Deep glabellar furrows
- Gonatoparian suture
- Eyes small, but prominent
- Cephalon semicircular
- Cranidium large; librigena also moderately broad
- Fixigena constricted between eye & glabella

Lichas (O. LICHIDA)

- Eyes small
- 11 thoracic segments
- Longitudinal glabellar furrows create a composite glabellar body by isolating the glabellar lobes
- Opisthoparian suture
- Genal angle extended into curved spine
- Pleural termination spinose and unreflexed
- Pygidium plain, spinose

1 cm

Peronopsis (O. AGNOSTIDA)

- Tapering lobate glabella ends short of anterior
- Axial ring wide
- Pleura very constricted
- Note diminutive size and isopygous
- 2 thoracic segments
- Plain, tapering axial area of pygidium

1 mm

FIGURE 3.9. *Calymene*, *Lichas* and *Peronopsis*.

3.8 Geological history and stratigraphic value

The Cambrian trilobite faunas underwent a crisis at the end of that period, so few of the old stocks made it into the Ordovician. There was a brief period of experimentation, when short-lived groups attempted to fill the ecological gaps left behind by these extinctions, but it was in the late Tremadoc that the major and 'typical' Ordovician groups became established.

After this there seems to have been little further development in the overall design of trilobites, with only minor variations on the basic themes stated in the early Ordovician. Those which survived that period were long-ranged, and many went on through the Silurian and into the Devonian. But it was another major crisis, from the middle to the end of that period, which dealt the trilobites their mortal blow. A few managed to survive through the Carboniferous, but the last seem to have died out in the later Permian.

Rates of development were high enough in the Cambro-Ordovician for trilobites to be used as zone fossils for correlation. Their major disadvantage is that their occurrence is not worldwide, different populations being separated into distinct faunal 'provinces' which may prevent or make more troublesome the business of long-distance correlation.

But being dominantly shelf-dwelling, characteristic of the relatively shallow-water 'shelly facies', trilobites enable the dovetailing of these sediments with the

Trilobites 27

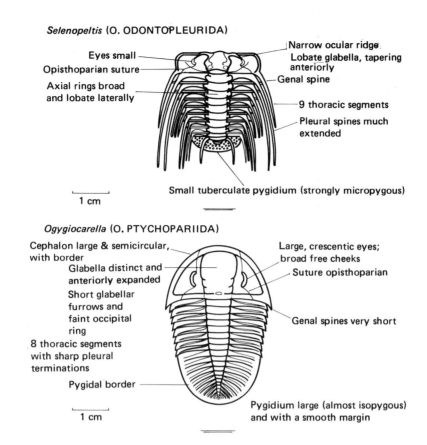

FIGURE 3.10. *Selenopeltis* and *Ogygiocarella*.

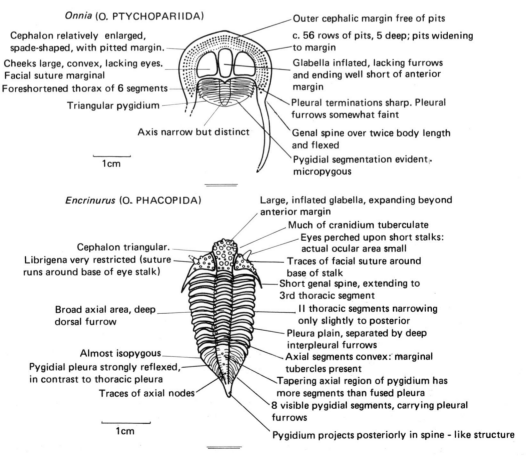

FIGURE 3.11. cf *Onnia* and *Encrinurus*.

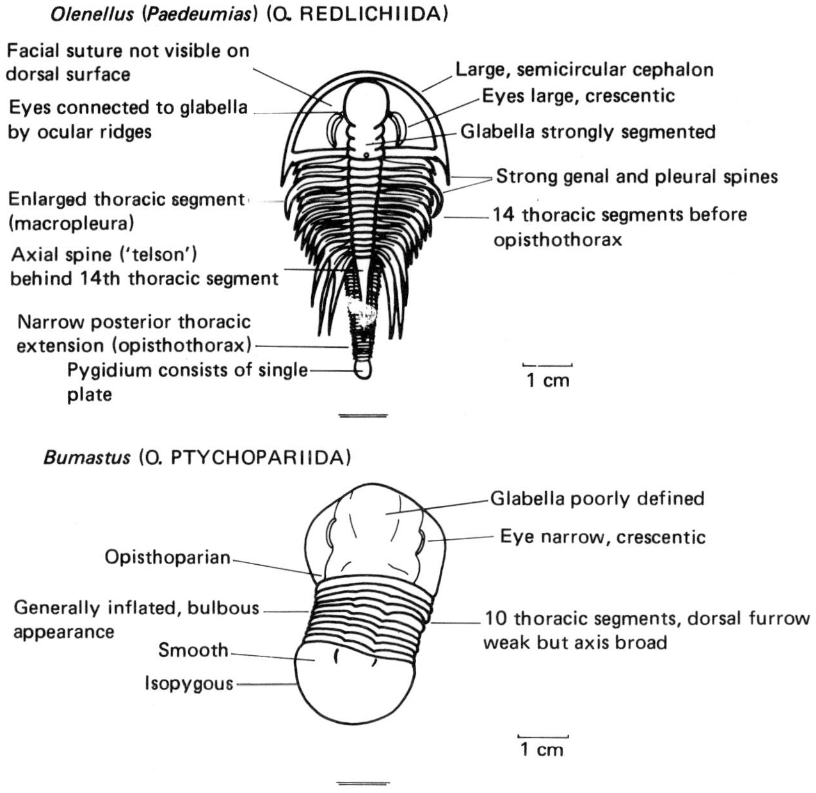

28 Palaeontology — An Introduction

Olenellus (Paedumias) (O. REDLICHIIDA)

- Facial suture not visible on dorsal surface
- Eyes connected to glabella by ocular ridges
- Enlarged thoracic segment (macropleura)
- Axial spine ('telson') behind 14th thoracic segment
- Narrow posterior thoracic extension (opisthothorax)
- Pygidium consists of single plate
- Large, semicircular cephalon
- Eyes large, crescentic
- Glabella strongly segmented
- Strong genal and pleural spines
- 14 thoracic segments before opisthothorax

1 cm

Bumastus (O. PTYCHOPARIIDA)

- Opisthoparian
- Generally inflated, bulbous appearance
- Smooth
- Isopygous
- Glabella poorly defined
- Eye narrow, crescentic
- 10 thoracic segments, dorsal furrow weak but axis broad

1 cm

FIGURE 3.12. *Olenellus* (*Paedumias*) and *Bumastus*.

basinal facies in which the graptolites occur. Graptolites rarely survive the more vigorous sedimentary environments of shelf seas.

The very provinciality of trilobite faunas may be put to use, however (Fig. 3.13 i). It had long been noticed that the trilobites in western Ireland and Scotland showed more resemblance to those of North America, Greenland and Newfoundland than to those found in England, Wales, Massachusetts, New Brunswick and south-east Ireland.

The reason for this can be found in plate-tectonic movements. In the Lower Palaeozoic, the present Atlantic Ocean had not formed, so North America and Greenland were joined to Scotland and N. Ireland. But these regions were widely separated by a huge ocean from those places we now call England, Wales, Massachusetts and New Brunswick. This long-vanished ocean has been called **Iapetus**.

On either side of Iapetus, separate trilobite faunas developed which were at their most divergent in the Llanvirn. As the ocean began to close, in the same way that the Pacific Ocean is closing today, the faunas on its opposing shores came closer together. Their resemblance to one another increased accordingly as migration of larval forms became possible across the narrowing sea. This narrowing process continued until the two faunas were identical, and it led eventually to the complete closure of the seaway and — the Caledonian Orogeny.

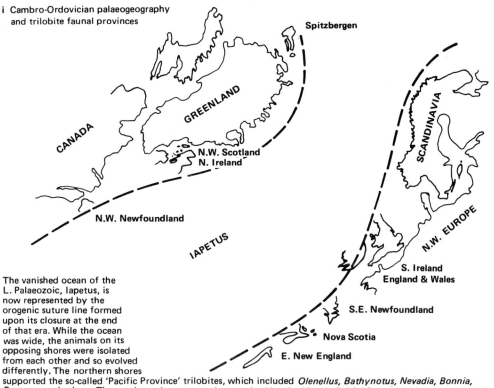

i Cambro-Ordovician palaeogeography and trilobite faunal provinces

The vanished ocean of the L. Palaeozoic, Iapetus, is now represented by the orogenic suture line formed upon its closure at the end of that era. While the ocean was wide, the animals on its opposing shores were isolated from each other and so evolved differently. The northern shores supported the so-called 'Pacific Province' trilobites, which included *Olenellus, Bathynotus, Nevadia, Bonnia, Protypus* and others. The southern shores were characterized by the 'Acado-Baltic' fauna, which included *Callavia, Holmia, Kjerulfia* and *Strenuella*. These faunas were established in the L. Cambrian. By the L. Ordovician, the northern shores supported a 'Bathyurid' fauna, while the southern margin was populated by the 'Selenopeltid' fauna. Other fossil groups, such as brachiopods, also reflect the dichotomy, but through the Ordovician the ocean narrowed — much as the Pacific Ocean is narrowing today — by the consumption of ocean floor beneath adjacent continents. As this process continued, cross-breeding and migration became possible, and the two faunal provinces gradually lost their independent identity.

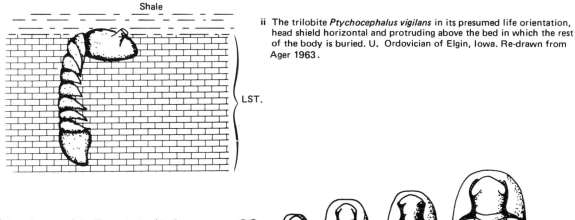

ii The trilobite *Ptychocephalus vigilans* in its presumed life orientation, head shield horizontal and protruding above the bed in which the rest of the body is buried. U. Ordovician of Elgin, Iowa. Re-drawn from Ager 1963.

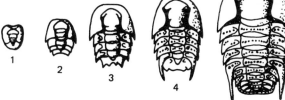

iii Larval stages of the Tremadocian (Ord) trilobite *Shumardia pusilla*. Stage 1, protaspis; 2-4, meraspis. 5, holaspis.

Figure 3.13.

4.
Graptolites

Phylum Hemichordata Subphylum Protochordata Class Graptolithina
Orders { Dendroidea (U. Camb.–L. Carb.)
        { Graptoloidea (L. Ord.–L. Dev.)

4.1 Introduction

In the dark, shaly sequences of the Lower Palaeozoic, you stand a reasonably good chance of finding graptolites. Their name could be translated as 'written on stone', reflecting their appearance of being inscribed on the bedding plane. Most of the graptolites you will see look like one or more hacksaw blades, with rather little morphological detail. They look insignificant, yet they can be used to subdivide and correlate the entire Ordovician and Silurian Systems. They are among the most useful tools for relative dating known to geology, and being members of the Phylum Hemichordata they are distantly related to all vertebrate animals — and that includes us!

Although themselves long extinct, graptolites have living relatives which help us to imagine what sort of animal created this skeleton. Most of the preservation seen in ordinary specimens is poor, and allows only the basic outline of morphology to be discerned. This is sufficient for the purposes of stratigraphy; but for detailed examination of their structure, use has been made of specimens so perfectly preserved that they are unflattened, still composed of their original skeletal material and even (in rare cases) still flexible.

Such superb fossils, under the correct chemical treatment, can be made to transmit light and reveal the minutest details of their construction. And recent work with the electron microscope has revitalized and redoubled the pace of graptolite palaeontology.

Graptolites come in two varieties: **dendroids** and **graptoloids**. Dendroids are less important stratigraphically and only occur locally, since in life they were attached and grew like small bushes on the sea floor. They are of great significance to the palaeontologist, however, who wants to work out the derivation and relationships of the graptolites as a whole. The dendroids first appeared in the Upper Cambrian, and gave rise to the graptoloids.

Graptoloids have smaller and much simpler skeletons than the bushy dendroids. They were free-floating (possibly free-swimming) and came to dominate the plankton throughout the Ordovician and Silurian. They ranged freely over very large expanses of ocean and occur widely in the deep-water sediments of those periods. They also lived in shallow waters, but were not preserved there because of the turbulence and the action of burrowing and browsing organisms which together destroyed most of the skeletons.

The immense practical use of graptolites adds to the strange fascination of an intricate form, once produced by a superbly successful group of creatures — creatures at whose true identity we can, alas, only guess.

4.2 Morphology

The skeleton of a graptolite is known as the **rhabdosome.** This may often be divisible into separate stick-like units called **stipes**. Each stipe is basically a hollow tube along which many small cups called **thecae** are connected. It is these which give rise to the serrated effect along the edges of specimens, and in each theca there once lived a separate graptolite animal, referred to as the **zooid**.

So a complete graptolite consisted of many individuals linked together by a **common canal**: hence, they were **colonial** animals. Coloniality is not a feature restricted to the graptolites, but is found throughout the lower invertebrate phyla, most notably in the corals (Ch. 8).

Growth of a graptolite colony proceeded from a primary individual which lived in the **sicula**, a conical cup which may, at its apex, be seen to bear a long thread or **nema**. These basic features are common to both orders of graptolite, but their more detailed morphology must be treated separately.

4.2.1 Dendroids

The growth of a dendroid graptolite began with the

attachment of the sicula to the sea floor. The cup opened upwards, and the whole colony developed from it, standing firm upon a holdfast which became much thickened for added strength.

The development of the rhabdosome proceeded by the growth of many branching stipes which were often cross-linked by transerve bars called **dissepiments**. Unlike the simpler graptoloids, the thecae of dendroids arose in groups, and each member of the group had a distinctive shape and, therefore, also a different function.

Thecal growth took place like this. From the sicula, a large cup called a **stolotheca** grew out (Fig. 4.1), and when this zooid budded, it gave rise to two new individuals. One was a large **autotheca** and the other a smaller **bitheca**. The stolotheca then grew ahead and later produced the next pair of buds, and so on. When the stipe branched, the stolotheca would split in two and each new daughter stolotheca would diverge to carry on budding in the normal way.

Autothecae and bithecae were therefore linked by the stolotheca which gave rise to them. Down the length of this stolotheca, and extending to the base of each daughter theca, ran a tubular thread called the **stolon**. This is believed to have been some kind of rudimentary nerve cord, and it is this feature which places the graptolites in the Hemichordata.

What, then, was the function of the autothecae and bithecae? It may have been that the large autotheca contained a female zooid and the bitheca a male. Such two-fold division of structure is usually sexual in origin (as it is in our own species) and is called **sexual dimorphism** (see Ch. 6, II).

While most dendroids were (as their name suggests) tree-like, growing attached to the sea bed, some became free-floating. These are characterized by a

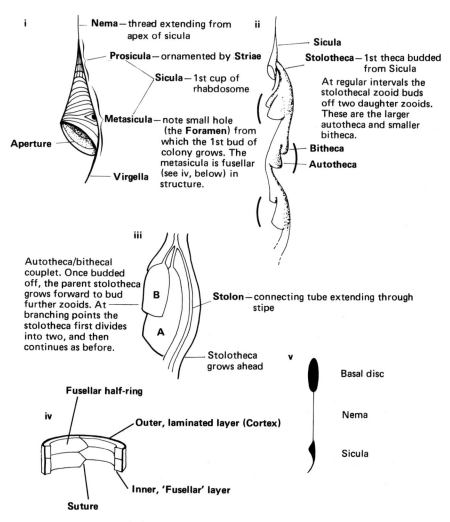

FIGURE 4.1. (i) Structure of the sicula. (ii) Budding process in dendroid graptolites. (iii) As for (ii), cut longitudinally to show relationship between thecae and the stolon. (iv) Graptolite periderm. Arrangement of fusellar half-rings, suture (one shown) and cortex. (v) Free-floating dendroids such as *Dictyonema* may have hung from a flotational structure by the nema. This may have been a gas-filled sac, or a vaned structure rather like dart-fletching. Diagrams all greatly enlarged and not to scale.

more delicate rhabdosome, and since they lost the ability to attach by their bases, they probably hung downwards by a thread from some floating object. In this way they became more widely distributed, and in life their thecae pointed downwards.

Dictyonema (Fig. 4.5), which is, perhaps, the most famous dendroid, was of this kind. The branching of its stipes is more regular than in the bushy, attached types, though the basic dendroid pattern of branching stipes with dissepiments remains unaltered in principle. *Dictyonema* is preserved flattened upon bedding planes, but in life it was a three-dimensional, conical structure.

As the planktic mode of life was adopted by more and more forms, certain trends in the evolution of their rhabdosomes are seen to have operated. Regularity of branching improved; the number of branches decreased, and in a haphazard and incidental way, this stipe reduction through time also took place in the dendroids' more successful derivations, the graptoloids.

4.2.2 Graptoloids

Graptoloids are made up, in most cases, of between one and four main stipes, although in the earliest examples the secondary branches of these may create an appearance of complexity. All the thecae on any one rhabdosome were of similar shape, and their microstructural details, discussed below, were different from dendroids. Graptoloids had no preservable stolon, although there was probably some similar structure passing down the common canal. Graptoloids never possessed dissepiments.

The graptoloid sicula was similar to that of dendroids. It was, on average, a little over 1.0 mm long and was extended apically into a nema. At the aperture, a projecting spine was developed, known as the **virgella**.

Upon the side of the sicula there grew an initial bud from which the first theca developed. This maintained contact with the sicular zooid via a hole (Fig. 4.1 i) and with all subsequent thecae by the common canal. The shapes of thecae vary enormously from group to group (Fig. 4.2), and these features are used to differentiate stratigraphically useful species. The taxonomy of the graptolites rests partly on this, partly upon the number of stipes present and partly upon the attitude of the stipes relative to the sicula (Fig. 4.2 iii).

All non-scandent rhabdosomes have stipes with thecae on one side only. They may commonly consist

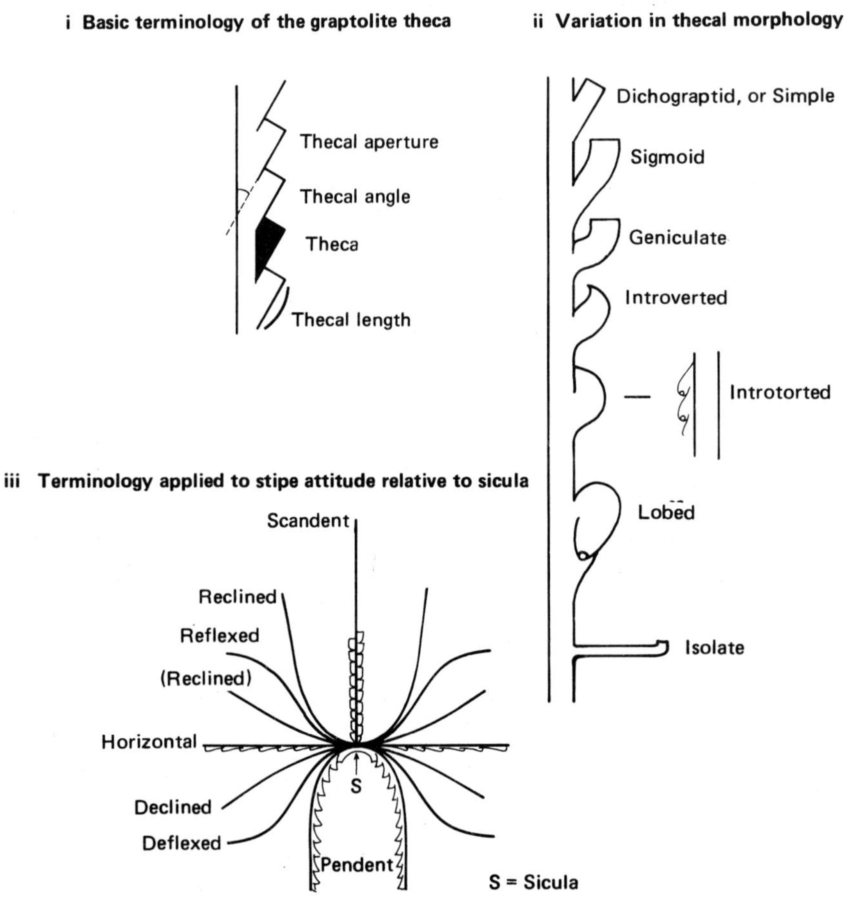

i Basic terminology of the graptolite theca

- Thecal aperture
- Thecal angle
- Theca
- Thecal length

ii Variation in thecal morphology

- Dichograptid, or Simple
- Sigmoid
- Geniculate
- Introverted
- Introtorted
- Lobed
- Isolate

iii Terminology applied to stipe attitude relative to sicula

- Scandent
- Reclined
- Reflexed
- (Reclined)
- Horizontal
- Declined
- Deflexed
- Pendent

S = Sicula

FIGURE 4.2. The graptolite vocabulary. Note, from (iii), that rhabdosomes may assume any position between pendent and scandent. Scandent graptolites may consist of one row of thecae (uniserial) as in the Monograptidae, two rows of thecae as above (biserial), typical of the Diplograptidae, or even three or four (triserial, quadriserial).

of two stipes (**biramous**) or four (**quadriramous**). **Scandent** rhabdosomes may form from the fusion of two stipes and so have two rows of thecae (**biserial**), four stipes and have four rows (**quadriserial**) or may consist of only one stipe growing upwards (**uniserial**), as in the monograptids (see Table 4.1).

Biserial graptolites may be subdivided upon the exact manner in which fusion took place (Fig. 4.3 v, vi). Those whose scandent stipes fused back-to-back shared one common canal. But members of the Glossograptidae fused side-to-side, like scissor blades. Each stipe, therefore, had its own common canal.

A useful method of determining graptolite phylogeny, i.e. of finding out which groups of graptoloids are most closely related to each other, is to examine the early stages of growth. The styles in which early thecae developed from the sicula to create the correct number and attitude of stipes, divide graptoloids into distinct types which are assumed to be genetically separate stocks (Fig. 4.3 i–iv). All the widely varying graptolite rhabdosomes are produced from the single initial bud upon the side of the sicula by these different modes of branching.

The morphological description of the many dendroid and graptoloid families is given in Figs. 4.5–4.11 and summarized for the graptoloids in Table 4.2 for ease of reference.

4.3 The classification of the graptoloids

Genera in graptoloids tend to be recognized by features of the entire rhabdosome, while species are distinguished by the different shapes of the thecae. This is by no means a hard and fast rule, but it remains broadly true. Development in new techniques of preparation and in microscopic examination is continually turning up new characters useful for classification, but this sort of detailed taxonomy creates problems for the geologist who would like to be able to distinguish zone fossils in the field — preferably without the use of an electron microscope! As the subject advances, however, such high levels of specialization become inevitable.

The problem is that morphological characters alone may evolve more than once in different stocks, due to a process called **convergent evolution** (see Ch. 12). Species or genera erected on these characters could therefore contain graptolites from different lineages, i.e. they would be **polyphyletic**. Since we want our classification to reflect the evolution of genetic groups, this is undesirable.

We would therefore aim for a classification based on large numbers of characters, including such minute details as early thecal development. In this we are hampered by the persistently poor preservation of

TABLE 4.1. *Nomenclature of Graptolite Rhabdosomes (S = sicula)*

Scandent	Uniserial (one row of thecae)	eg. *Monograptus*
	Biserial (two rows of thecae)	eg. *Diplograptus*
	Quadriserial (four rows of thecae)	eg. *Phyllograptus*
'Non-scandent' (i.e. pendent to reflexed)	Biramous (two stipes)	eg. *Didymograptus*
	Quadriramous (four stipes)	eg. *Tetragraptus*

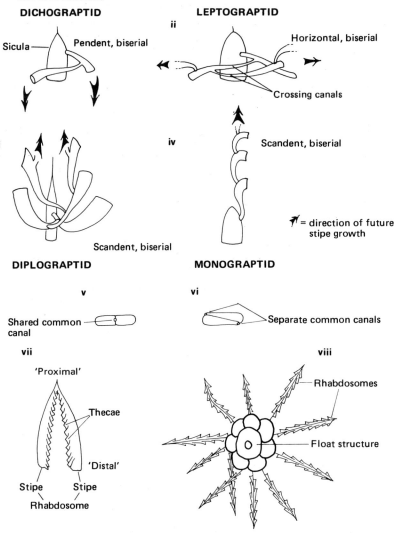

FIGURE 4.3. (i)–(iv) Cartoons to show the four basic styles of initial budding in graptolites. (v), (vi) Transverse sections of two scandent, biserial forms. In (v) the stipes fuse back-to-back and the common canal is shared. In (vi) the stipes fuse side-to-side like scissor blades, each preserving its common canal. (vii) Basic terminology applied to a pendent graptolite. (viii) A 'synrhabdosome': many rhabdosomes clustered around a common centre. Diagrams (i)–(vi) much enlarged and not to scale. Diagrams (vii) and (viii) not to scale.

such delicate fossils. As a result, the classification shown in Table 4.2 is at best a compromise solution.

4.4 The structure of the graptolite periderm

For a long time it was thought that graptolites were made of chitin, but we now know that their periderm was constructed of a fibrous scleroprotein (see Ch. 1). The feeling is that this was probably **collagen**, a substance which in modern animals forms tissues of high tensile strength (tendons, for example).

In both dendroids and graptoloids the periderm was bi-layered. The inner one, called the **fusellar layer**, was constructed in growth increments which themselves comprised two semicircular units (Fig. 4.1 iv). These joined along a zig-zag suture, and each increment may have represented one day's growth.

Over the exterior, the second peridermal coat was laid down. This was composed of surface-parallel sheets plastered over the fusellae, rather like rendering over bricks. It is known as **cortical tissue**, and how it was deposited is something of a mystery, because its existence seems to imply the presence of living tissue beyond the edges of the thecae.

The use of very powerful microscopes to investigate the consistency of the two different types of periderm has revealed that the inner, fusellar tissue was made from scleroprotein bundles in a rather loose arrangement. The cortical tissue, by contrast, was made up of coarser fibres, but lying parallel to one another. It therefore conveyed structural strength to the lighter fusellar material, especially when present in great thickness.

This serves to explain the differences between

TABLE 4.2. *Graptoloid classification*

Suborder	Family	Typical example	Description
Didymograptina	Dichograptidae L. Ord.–M. Ord.		Bilateral symmetrical, lateral or dichotomous branching. Stipes declined or horizontal, uniserial (usually). May be biserial or quadriserial eg. *Tetragraptus* (Arenig) *Didymograptus* (Llanvirn/Llandeilo) *Phyllograptus* (Arenig–Llanvirn)
Didymograptina	Nemagraptidae M.–U. Ord.		Biramous, slender flexuous stipes horizontal or reflexed in most cases. Sigmoid thecae. eg. *Leptograptus* – marker for basal Bala.
Didymograptina	Dicranograptidae M.–U. Ord.		Uniserial or both uni- and biserial – eg., *Dicranograptus*, which is biserial proximally and uniserial/biramous distally. Found mainly in the Llanvirn.
Glossograptina	Glossograptidae L.–M Ord.		Scandent biserial, with stipes fused side-to-side in scissor-blade fashion. Spines are common. eg. *Cryptograptus* (spines proximally), *Glossograptus*, with spines at apertures of all thecae.
Glossograptina	Cryptograptidae L.–M. Ord.		
Diplograptina	Diplograptidae L. Ord.–L. Sil.		Scandent, biserial: stipes mostly straight. Sigmoid thecae. – eg., *Climacograptus*. This form has geniculate thecae, alternate upon the rhabdosome. *Orthograptus* may occur in synrhabdosomes.
Diplograptina	Lasiograptidae M. Ord.–U. Ord.		Scandent, biserial: periderm reduced to lacy network. Possibly a tactic to reduce specific gravity & enhance buoyancy – eg., *Lasiograptus* (U. Arenig – Bala)
Diplograptina	Retiolitidae M.Ord.–U. Sil.		Scandent, biserial: as above, periderm reduced to a reticulate network. eg. *Retiolites* (L. Sil–M. Sil.)
Diplograptina	Dimorphograptidae L. Sil.		Scandent forms with compound morphology eg. *Dimorphgraptus* (L. Sil.) which is characteristically scandent uniserial at base, becoming biserial distally. Marker for Ord/Sil boundary
Monograptina	Monograptidae Sil.		Scandent uniramous graptolites, straight and curved. Typical of the Silurian, which is largely zoned on monograptids. During their broad evolution, there was a tendency for thecae to become more isolated
Monograptina	Cyrtograptidae Sil.		Uniramous, curved and often spiral forms with side-branches called cladia eg., *Cyrtograptus* (Wenlock).

dendroid and graptoloid periderm. The dendroids, being fixed to the sea bed, required structural stability and so display very large quantities of secondary, cortical tissue: especially around the much-thickened holdfast area. With the adoption of the floating mode of life, this dense material was more of a hindrance than a help, and so we see a progressive reduction in cortex towards the appearance of the true graptoloids, where it exists only as a thin outer covering.

4.5 Graptolite affinities. What was the graptolite animal?

Sometimes the fossil remains of extinct groups of animals leave no doubt as to their correct biological identity. The trilobites, for instance, are obviously arthropods, like insects or crustaceans. However, there are other groups which may be far more difficult to assign, since they may bear no immediately striking resemblance to well-known living creatures. If this difficulty persists, such groups are then referred to as 'problematica' until such time as the balance of opinion is so strongly in favour of a particular theory that they can be moved to a definite position in the taxonomic system.

In later chapters we shall come across groups which are still problematical, but the graptolites are a group whose affinities have been settled only in the last forty years. Such settlements are rarely unanimous, but they are usually made by reference to the fossil material itself and to the structure of some living group which careful study has shown it to resemble. Such a revelation has, more recently, been accepted with regard to the stromatoporoids (see Ch. 8) and we live in hope that all problematical groups will one day be resolved in this way. Realistically, though, it seems quite likely that some of them will remain controversial for ever.

Rhabdopleura is a colonial animal which lives at the bottom of the sea, mostly in the southern hemisphere. It consists (Fig. 4.4) of many tubular sections branching from a main tube which creeps along the sea bed and has at its end the closed terminal bud which from time to time gives rise to the daughter zooids forming

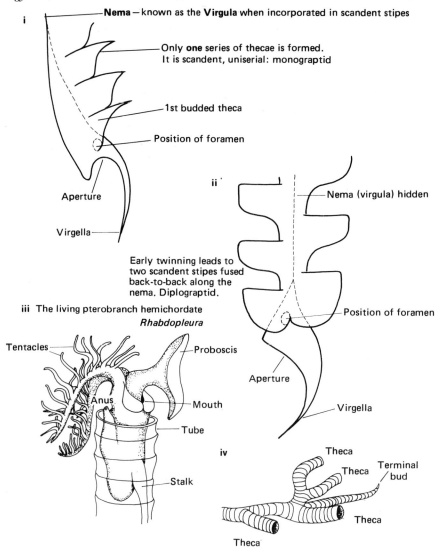

FIGURE 4.4. (i), (ii) Thecal addition in monograptid and dichograptid respectively. (iii) Zooid of a living relative of the graptolites, *Rhabdopleura*. The tubes of the colony (iv) are constructed of fusellar rings in a very similar manner to that of graptolites. All diagrams much enlarged and not to scale.

the branches. It is linked to all these by a stolon system; the skeleton is made of fusellar scleroprotein, each tube bearing two distinctive zig-zag sutures where the fusellae come together. You can see that the skeletal similarities between *Rhabdopleura* and graptolites are very strong indeed.

The zooid of *Rhabdopleura* (Fig. 4.4) consists of a pair of feathery food-gathering arms which extend out from the end of the tube. Each zooid is retractable, because the body is attached to the stolon by a contractile stalk. If the affinity between these modern hemichordates and the graptolites is correct, the graptolite animal may have looked very like this. Today the graptolites are designated as a class within the Phylum Hemichordata.

4.6 Mode of life

The development of dendroids towards a floating mode of life and the exclusive adoption of this habit by graptoloids is reflected in the changing peridermal structure as suggested above. Other aspects of colonial design and peridermal reduction can be viewed in a similar light, tending to improve flotation by lowering specific gravity or increasing surface area relative to mass.

Dictyonema, a floating dendroid, may have been attached to floating material such as seaweed; indeed, many graptolites may have taken up this 'hitch-hiking' mode of existence. Some fossils of *Dictyonema*, however, have revealed a black disc at the end of the nema. This could have been a sac filled with fatty material or gas, but it may also have been a vaned structure, like the flights of a dart (Fig. 4.1).

Some dichograptids appear to have grown not as isolated rhabdosomes but in aggregations of colonies around a central body. These are called **synrhabdosomes**, and were probably a kind of supercolony (Fig.

4.3). The central (?flotational) disc was surrounded by capsules containing siculae; these may have been awaiting release either to become attached to the parent supercolony or to go their own way, maybe to found new synrhabdosomes.

Although the spines of trilobites probably had little practical effect upon their buoyancy (except, possibly, in larval stages), the spines of graptolites are almost certain to have had such a function. The principle is the same as that adopted in the dispersal mechanisms of seeds — such as those of dandelion or thistle — which improve buoyancy.

The same effect could have been achieved by reducing the mass directly rather than adding to the surface area. This was done by the Lasiograptidae and Retiolitidae (Table 4.1), two very successful graptoloid families. The representatives of these groups reduced their periderm to a lacy reticulate network, and many also produced long spines either basally or covering the entire rhabdosome.

The Cyrtograptidae are characteristically curved — as are certain species of monograptid (Fig. 4.11) — and helical coiling is not uncommon. This form, too, would have acted as a flotation aid; the analogy in this case would be with the seeds of the sycamore, with their propeller wing to prolong the falling process and so aid dispersal.

Lastly, it has been suggested quite recently that graptolites may not have been mere passive floaters, but active swimmers: not planktic but nektic, propelled along by the strength of the zooids' feeding currents. The reduction in stipe number throughout the evolution of the Graptoloidea could be seen in the context of this theory, as a means to increasing the efficiency of this process. But this hypothesis is one which remains controversial.

4.7 Geological history and stratigraphic value

The Tremadoc (L. Ord.) saw the development of dendroids with a free-floating habit, such as *Dictyonema*: and with them came the important inter-

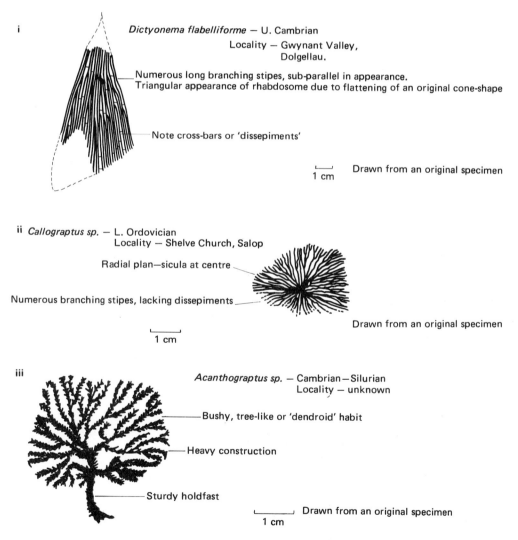

FIGURE 4.5. Dendroid graptolites.

mediate forms which demonstrate the derivation of the graptoloids from the dendroids, the Anisograptidae. No graptoloids are known from the Tremadoc.

The base of the Arenig, however, is defined by the appearance of a four-armed dichograptid graptoloid, *Tetragraptus* (Fig. 4.6 ii). Dichograptids dominate the fauna of this series and serve to distinguish it from all subsequent series of the Ordovician, which were all dominated by diplograptids of one sort or another. Recognition of this 'faunal type' is the first step in establishing the relative age of Ordovician and Silurian rocks. The species-level identification may then define the sediments to a precise zone. Take a look at Figs. 4.5–4.9 and see if you can pick out the significant limits of certain graptoloid families — such as the Monograptidae, for instance. You will find a more comprehensive survey of ranges in Table 4.2.

The zonation established upon graptolite species is good, but on the whole the species were rather long-lived. Thus in Europe we have two Tremadocian zones, eleven in the rest of the Ordovician, thirty-two Silurian ones and three in the Devonian. This gives us an approximate figure of about one million years per zone on average. The much shorter-lived ammonite species of the Jurassic can subdivide that period into an average chron (a chron is the unit of time during which the sediments of a zone are deposited) of

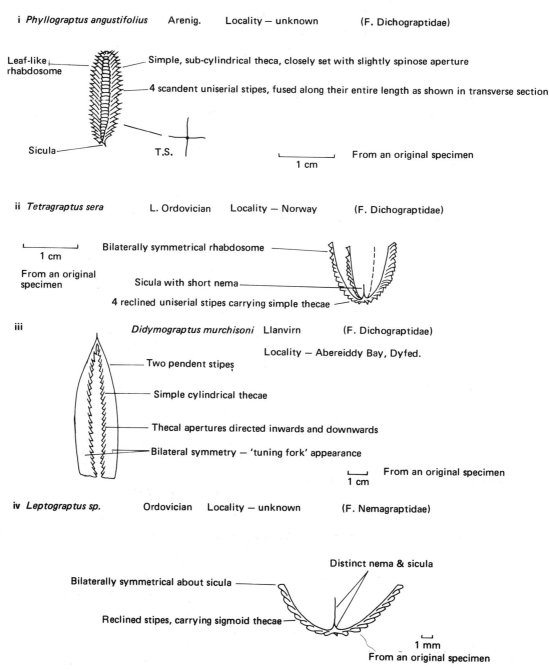

FIGURE 4.6. Dichograptid and Nemagraptid graptolites.

770,000 years. The shortest Jurassic chron may be as little as 400,000 years.

So graptolites are very good stratigraphic tools. But what — apart from their rather stable species — are their shortcomings? Geologists need to know what these are in order to know how far to trust the evidence they provide. Well, we have mentioned one of them already; their poor preservation record in shallow-water sediments. This means that, as fossils, graptolites are not as facies-independent as the living creature originally was. This is not unusual, and the ammonites were not totally uninfluenced by facies; but at least their more robust shells could survive burial in a wide range of possible environments.

Another drawback with graptolites, noticed with the extension of Palaeozoic stratigraphy over the globe, is their tendency for **provincialism**. These provinces were understandably very large, but that they should exist at all is something of a surprise. What prevented them, as floating animals, from attaining worldwide distribution?

It is believed that the reason for this was large-scale oceanic circulation. The Atlantic is at present divided into two great circulations or **gyres**, one in the northern hemisphere and another in the south. They are of similar scale to the graptolite provinces, and similar gyres may have served to contain distinct populations of plankton within their broad, pan-oceanic sweep.

The evolution of the graptolites is characterized by

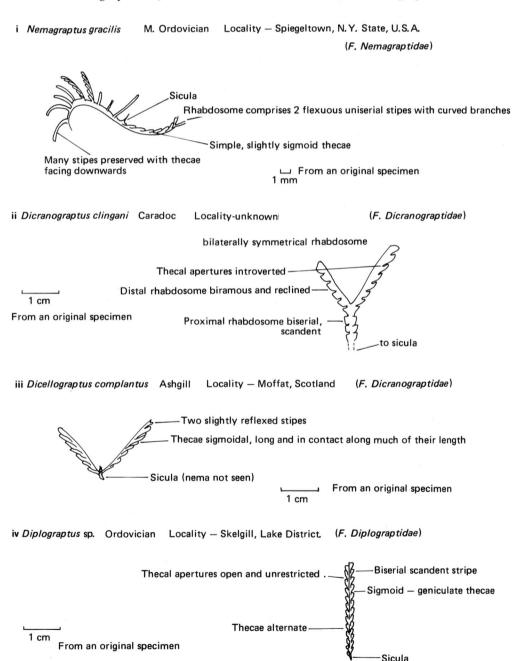

FIGURE 4.7. Nemagraptid, Dicranograptid and Diplograptid graptolites.

few genuine trends. There is a tendency, as has been mentioned, for the number of stipes to be reduced with time, but this appears to be incidental and discontinuous. Nevertheless, the graptolites have been cited in the past as exemplifying gradualistic evolution.

One hypothesis in particular held that the scandent biserial graptolites evolved from pendent, biramous types by a series of progressive changes. It envisaged the pendent stipes lifting themselves to the horizontal as evolution proceeded, and eventually becoming upraised. The climax of this supposed progression was reached when the stipes became vertical, were aligned back-to-back and were fused together. The genus *Dicranograptus* (Fig. 4.7 ii), a Y-shaped form which is scandent and biserial at its proximal end but diverges to become biramous distally, was cited as a half-way stage in this process, like a zipper half done up.

This hypothesis was widely accepted until it was realized that the first diplograptid (*Glyptograptus dentatus*) occurs in the Upper Arenig, millions of years *before* the appearance of *Dicranograptus* in the Llandeilo. Moreover, work on the early growth stages of dicho- and diplograptids has revealed that they are fundamentally different (Fig. 4.3). The inference to be drawn from this is that dichograptids and diplograptids are drawn from different genetic stocks, and that therefore the evolution of one into the other by a simple raising of the stipes is very unlikely.

On the whole, the major changes in the evolution of graptolites took place suddenly, the new forms appearing without apparent ancestry. Between these major leaps, however, minor design refinements appear to have taken place gradationally. This kind of evolutionary pattern, involving major leaps and minor intervening alterations upon each revolutionary new development, is in keeping with modern ideas upon the nature of evolutionary change and biological innovation. For a discussion of these ideas see Chapter 12.

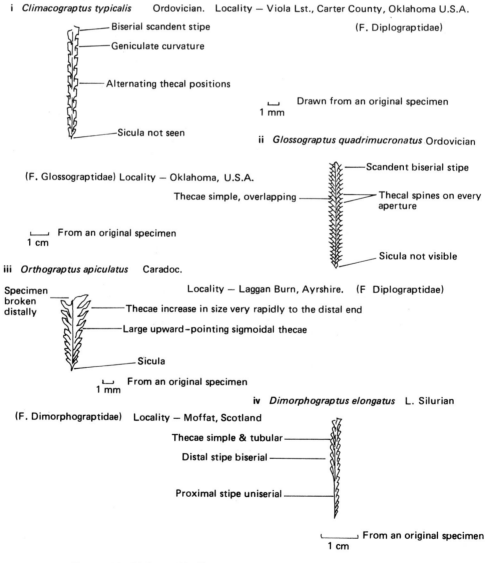

FIGURE 4.8. Diplograptid, Glossograptid and Dimorphograptid graptolites.

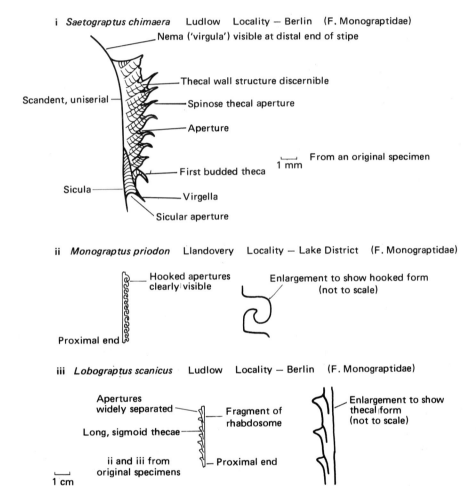

FIGURE 4.9. Monograptid graptolites.

42 *Palaeontology — An Introduction*

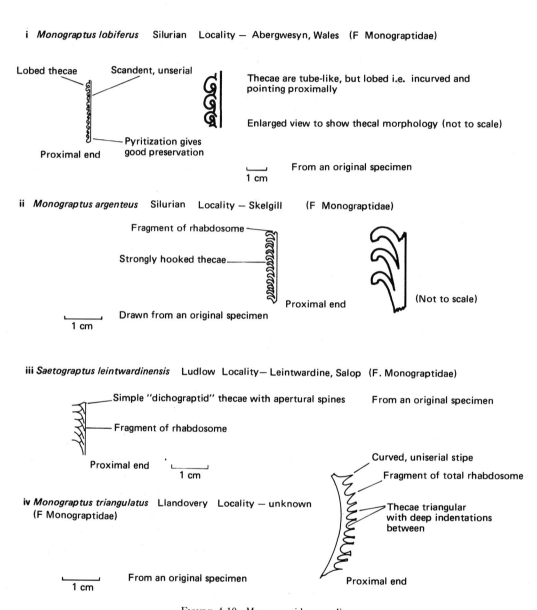

Figure 4.10. Monograptid graptolites.

Graptolites

i *Cyrtograptus murchisoni* Wenlock Locality — unknown (F. Cyrtograptidae)

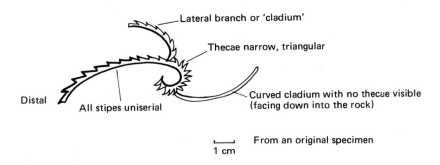

ii *Rastrites maximus* Llandovery Locality — Moffat, Scotland (F. Monograptidae)

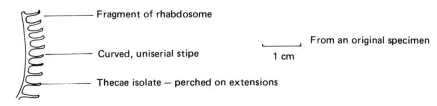

iii *Spirograptus spiralis* Llandovery Locality — Bohemia (F. Monograptidae)

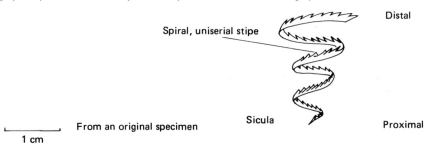

FIGURE 4.11. Cyrtograptid and Monograptid graptolites.

5.
Brachiopods

Phylum Brachiopoda ⎯⎯⟨ Class Articulata / Class Inarticulata ⎯ (L. Camb.–Rec.)

5.1 Introduction

Because brachiopods are very rarely found in British coastal waters we have no common name for them in English. They are sometimes referred to as 'lamp shells', but unless we are familiar with Roman oil lamps (which they reputedly resemble) this name conveys little idea of their appearance. This is a shame, because the brachiopods are one of the most important fossil groups, very abundant in sediments of shallow seas and useful indicators of environment. Also, the existence of living representatives enables us to relate the fossil shells to the anatomy of the animal with some confidence.

Brachiopods are benthic marine invertebrates with a shell composed of two **valves** hinged together and made of calcite. In their basic form they resemble the bivalved molluscs (Ch. 6) such as cockles and mussels, with which we are all familiar. Nevertheless, this similarity is only superficial, for as well as having an entirely different anatomy, even their shells may be seen to be quite distinctive when one examines their symmetry (Fig. 5.1). In bivalves, the plane of symmetry lies *between* the two valves (i.e. the valves are mirror images). With brachiopods, the plane divides both valves into two.

During the long history of this phylum, the simple basic form of the brachiopod has undergone a myriad modifications to suit the demands of different environments and to allow these creatures to exploit ever more diverse modes of life (Table 5.1). It is their plasticity and evident adaptability which gives brachiopods their special fascination, because it has been achieved with a very simple and elegant basic design.

Most brachiopods live tethered to the sea floor by a structure called the **pedicle** (Fig. 5.2), filtering food particles from the currents which they pass continuously in and out of their slightly gaping shells. This passive life style extends also into their reproductive behaviour, which merely involves the release of eggs and sperm into the sea, trusting to luck for fertilization. Brachiopods tend to live in clusters, so this system, for all its lack of enthusiasm, is nevertheless quite efficient. In most species the sexes are separate, though **hermaphroditism** (each individual being both male and female) is known.

Following successful fertilization, the larval brachiopod spends a short while swimming freely, after which time it begins to search out a suitable attachment site where it will settle — usually for life.

5.2 Morphology and internal anatomy

The phylum is divided into two classes, the **articulates** and the less important **inarticulates**. Their names point to one major difference between them, namely the absence, in inarticulates, of **teeth** and **sockets** to fix the two valves together along the hinge. The two classes therefore have radically different systems of musculature. Other differences include shell composition; inarticulates are most commonly made of chitin interlayered with calcium phosphate, while articulates are calcitic. Also, the pedicle of inarticulates forms quite differently from that of the articulates, and may be muscular and contractile where that of articulates is inert.

We shall be considering the morphology of these two classes separately in this account. Nevertheless, it is important not to lose sight of the fundamental features which they have in common, and which unite them within the same phylum.

In both classes the valves are bilaterally symmetrical, and one of them is larger than the other. The large valve, which may only be slightly bigger than its counterpart, is called the **pedicle valve**, since in pedunculate forms (those bearing a pedicle) the pedicle emerges through it.

The smaller valve is known as the **brachial valve**, and in many species it bears on its internal surface two projections called **brachidia** (see below).

In other books you may also see the pedicle valve called the 'ventral' and the brachial valve 'dorsal'. It should be understood that these words do not refer to

TABLE 5.1. *Summary of Brachiopod Geological History*

System	Major events	Dominant groups	Subordinate groups	Extinctions
Quaternary		Terebratulida	Rhynchonellida Inarticulata	
Tertiary	Phylum assumes only minor status	Rhynchonellida Terebratulida		
Cretaceous		Rhynchonellida Terebratulida (great diversity displayed)		
Jurassic	Expansion of Rhynconellida and Terebratulida	Rhynchonellida Terebratulida		Strophomenida Spiriferida
Triassic	Appearance of Rhynchonellida & Terebratulida	Spiriferida	Rhynchonellida Terebratulida	
Permian	Mass extinctions all but obliterate the Brachiopoda	Productidina—only to become extinct by end. Spiriferida		Orthida, many Rhynchonellida & Terebratulida
Carboniferous	Radiation of the Productidina	Productidina Spiriferida	Rhynchonellida, Terebratulida	
Devonian	Many articulate orders become extinct	Spiriferida—esp. the Spiriferidina	Terebratulida first appear	Pentamerida Atrypidina & impunctate Orthida
Silurian	Further orders of articulates & inarticulates appear	Pentamerida—eg., *Pentamerus, Conchidium.* Spiriferida—esp. Atrypidina		
Ordovician	*Lingula* appears. Articulates undergo rapid evolution	Inarticulata and early articulates— Orthida	All remaining major groups (except Terebratulida) appear	
Cambrian		Inarticulata — eg., *Lingulella*	Small numbers of articulates eg. Orthida, Pentamerida	

the position which the shell adopts in life, but rather to its developmental origin within the larva. These terms are not used here, but the corresponding 'front and back' terms (anterior and posterior) are very convenient and are shown in Figs. 5.1 and 5.2.

5.2.1 Soft parts

The body of a brachiopod (Fig. 5.3) does not occupy all the space inside the shell. Most of this is taken up by a cavity called the **mantle cavity**, within which the **lophophore** (see below) is suspended. The mantle lines the internal surfaces of the valves and is responsible for secreting them. It is also a sensory organ. The cells around the edge of the mantle are sensitive to light, chemicals and touch, and stimulation of the sensitive area may cause the shell to snap shut.

At the posterior end of the shell the body itself resides, separated from the mantle cavity by the body wall and communicating with it by the mouth. Around this opening lie the two arms (**brachia**) of the lophophore. This is a feathery organ whose function is to create water currents, filter the water for food and act as a gill by taking up oxygen and releasing carbon dioxide. From the axis of each brachium extend many slender filaments clothed in millions of cilia and a

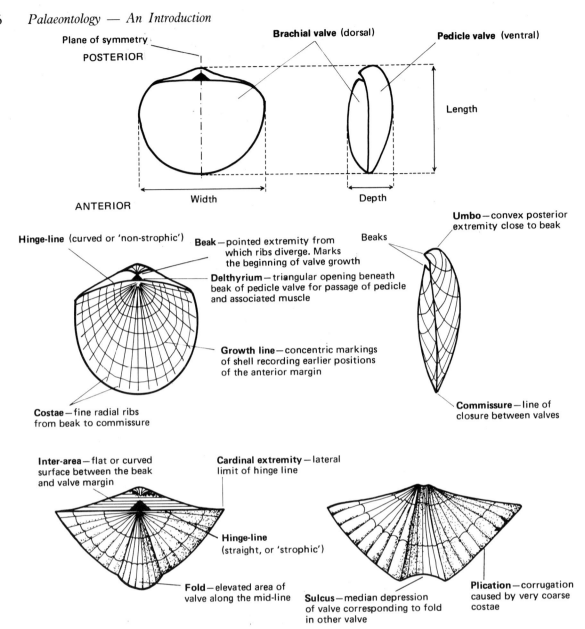

FIGURE 5.1. Basic external morphology and orientation of the shell in brachiopods (articulates).

sticky mucus. This entraps food particles which the cilia then waft back towards the axis where the **food groove** conveys them to the mouth.

The mouth of articulate brachiopods leads to a short, blind-ending gut. This means that faecal material must be stored and ejected via the mouth from time to time. Inarticulates have a much more convenient **anus** for this purpose. Metabolic wastes (which in humans are passed out in the urine) are removed from the body by a pair of 'kidneys' called **nephridia**. The nephridial pores also allow the eggs and sperm to escape into the outside world for fertilization.

Passing through the body are the muscles which open and close the shell, each function (in articulates) being performed by a separate set. Muscles only work by contraction, never by expansion. This means that the closing muscles (**adductors**) must be fixed to the anterior side of the hinge, while the opening muscles (**diductors**) must somehow act upon the opposite side.

As we shall see, the mechanical solutions to this requirement have been many and varied. A third set of muscles in pedunculate forms acts upon the pedicle and serves to change the brachiopod's position in the water. They are called **pedicle adjustor muscles**.

Mention has already been made of the fact that inarticulates have a different musculature (Fig. 5.2). Firstly, they lack diductors, and merely gape when the adductors relax. Secondly, to control the alignment of the valves relative to each other a set of **oblique muscles** is developed. Lastly, the pedicle itself may contain its own muscles.

Muscles leave distinct marks on the insides of shells at the points where they attach. These **muscle scars**

Brachiopods 47

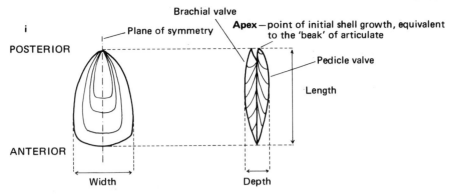

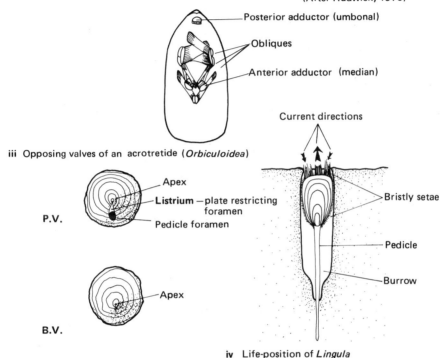

FIGURE 5.2. Internal and external morphology, orientation and life-position of inarticulate brachiopods.

allow us to reconstruct the lines of action of the muscles in fossil forms. While they are quite simple in articulates, the additional oblique muscles of inarticulates make their often indistinct scars rather more difficult to interpret.

5.2.2 Hard parts (articulates)

The terms used to describe the external morphology of articulate brachiopods are explained in Figs. 5.1, 5.4 and 5.5. Some of these are purely descriptive, but others need special discussion because of their bearing upon the animal's mode of life.

The pedicle emerges through a hole in the posterior region of the pedicle valve (Fig. 5.5). This usually takes the form of a triangular notch beneath the beak which widens toward the hinge line. Through this the round-sectioned pedicle (and, in some cases, certain muscles) must pass. They may not block the **delthyrium** completely, in which case shelly outgrowths may encroach so as to enclose the projecting organs more closely. These outgrowths are called **deltidial plates** and the hole which they define is then referred to as the **pedicle foramen**.

In some other brachiopods the reverse of this problem may have been encountered, namely, there may have been too little space allowed by the delthyrium for the passage of the pedicle and muscle. In these forms the hole may be enlarged by the creation of a similar notch in the brachial valve, directly opposite the delthyrium (Fig. 5.5 v). This is called the **notothyrium**, but its function is the same as that of the delthyrium which it enlarges.

Inside the shell are four groups of features which are especially important. These are (a) the **muscle scars**, (b) the **platforms** upon which muscles were sometimes

48   *Palaeontology — An Introduction*

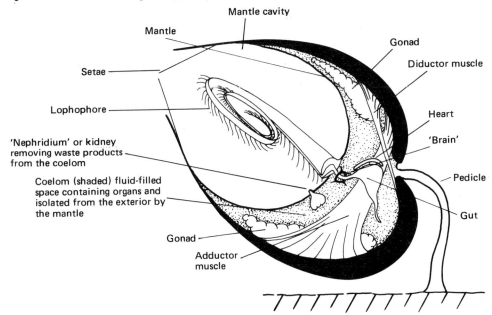

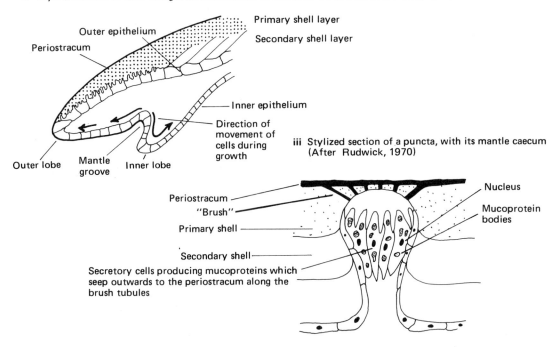

Figure 5.3. Brachiopod soft parts and their relationship with the shell.

perched, (c) the **teeth** (on the pedicle valve) and **sockets** (on the brachial valve) which lie on the hinge axis and (d) the **brachidia**, which suspend the lophophore from the brachial valve. Most are important in classification.

Muscle-scar shape, size and pattern (Fig. 5.5 iii) are important to note if visible. The scars shown in the diagram occur on the smooth internal surface of the valve, and are usually slightly depressed relative to their surroundings. In some brachiopods, notably the very bulbous kind, muscles would need to be very long to reach from the top of one valve to the bottom of the other, and since long muscles are inefficient, this is a problem.

Try opening a door by pulling on one end of a very long elastic band fixed to the door handle. You will find that the door opens, but that it will only do so after a delay; the initial force of your action is absorbed for a while in the elasticity of the band. Similarly, over-long muscles are sluggish.

To overcome this difficulty, some brachiopods introduce inelastic tendons into their muscles (Fig. 5.6 iii). This would be like opening the door using a short elastic band and a long piece of string tied to it.

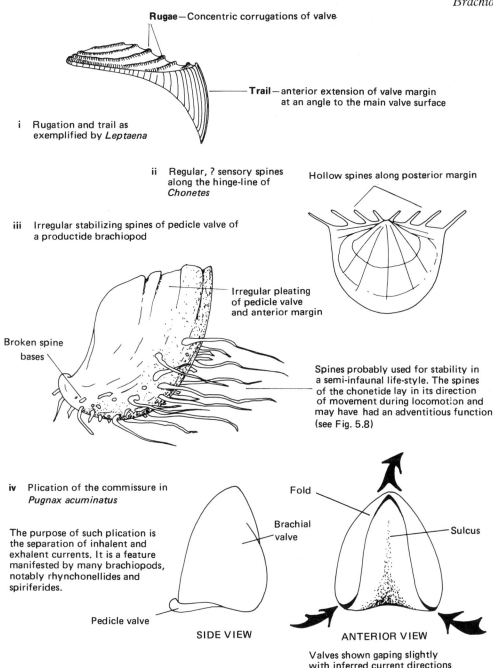

FIGURE 5.4. Aspects of gross external morphology in articulates. Not to any scale.

Others, unable to make tendonous muscles, raised their muscles on **platforms**. This would be like attaching our short rubber band to a long bracket projecting from, and firmly fixed to, the door. Muscle platforms are easily seen in pentameracean brachiopods (Fig. 5.5 iv), which actually get their name from the way these platforms appear to divide the interior space into five sectors.

Brachidia are very variable in design. Some may be short prongs (as seen in rhynchonellides) or loops, as seen in terebratulides (Figs. 5.5 i, 5.14). Most elaborate were the spiral brachidia of the spiriferides (Fig. 5.5 ii). These spires supported spiral lophophores which had an immense surface area and were able to divide up the mantle cavity into two distinct volumes — so creating currents more efficiently.

Teeth vary in their degree of development from species to species, but a discussion of these is best left until the mechanics of brachiopod hinges has been discussed (section 5.4).

5.2.3 Hard parts (inarticulates)

Lingula, the most well-known modern inarticulate and record-breaking genus for evolutionary conservat-

50   Palaeontology — An Introduction

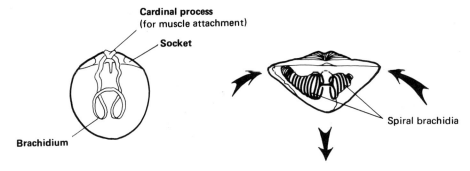

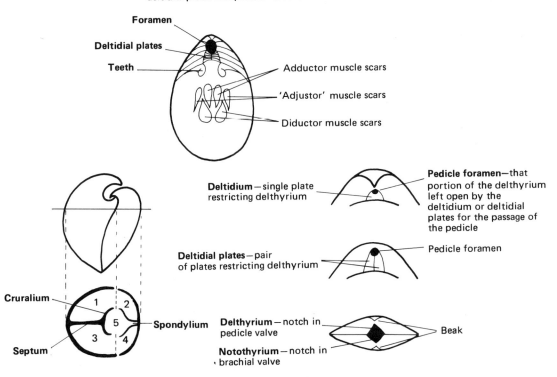

FIGURE 5.5. Internal skeletal features of articulates and delthyrial morphology.

ism, is by no means typical of all inarticulates. Those which, like *Crania*, lived cemented to hard surfaces had a tough calcareous shell whose fixed valve was almost obscured beneath the conical free valve. *Orbiculoidea*, a pedunculate form, had two conical valves of circular outline and a pedicle which emerged via a distinctive cleft or furrow in the pedicle valve (Fig. 5.2 iii).

Lingula itself had two almost identical valves. Both were gently convex and spade-shaped, made of chitin and calcium phosphate and reaching 1–2 cm in length. In modern representatives of the genus, the valves are translucent, and the delicate growth lines clearly visible. The pedicle emerges through a hole shared by both valves, and may be up to fifteen times the length of the shell. Its purpose is to anchor the shell in its vertical burrow (Fig. 5.2 iv).

The burrow opens by a narrow slit through which stiff bristles called **setae** project from the anterior mantle edge. They form themselves into three tubes: the outer two inhalent, and the central one exhalent.

Internal structures of inarticulates are rarely seen. There may be muscle platforms, but features of the inner surface are usually confined to faint and rather complicated muscle scars.

5.3 Shell growth, composition and structure

The secretion of the mature shell begins immediately after the larva settles. The mantle is responsible

i   Hinge types in articulates

Strophic **Non-strophic**

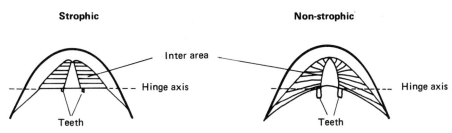

In strophic hinges, the hinge line is straight and coincides with the hinge axis. Non-strophic hinges have curved hinge lines and the hinge axis therefore only crosses the hinge line at two points. This system is inherently weaker, so teeth are often more prominently developed on non-strophic hinges.

ii Mechanical arrangements in articulate hinges. Hinge axes marked by concentric dot and circle. DM = diductor muscle. AM = adductor muscle. Shells drawn as though transparent. Re-drawn from Rudwick 1970.

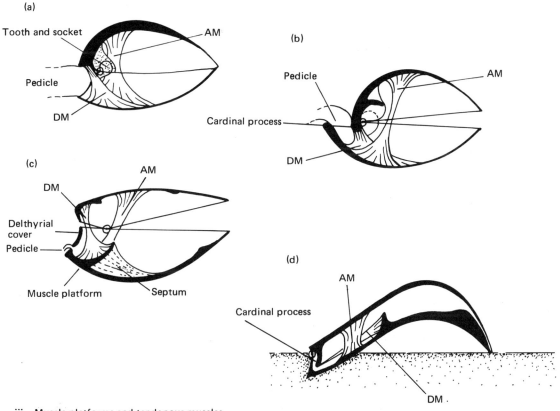

iii Muscle platforms and tendonous muscles

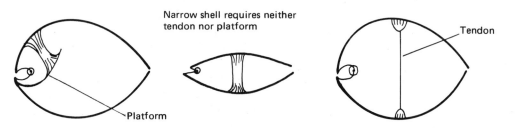

Long muscles are inefficient. In convex shells therefore the muscles are either raised up on platforms or else they are tendonous — i.e. consist of short muscle fibres attached to inelastic tendons. After Rudwick, 1970.

FIGURE 5.6. Muscles and hinges in articulate brachiopods.

for secreting the shell, and this process continues throughout the animal's life, so that the oldest shell near the umbones is also the thickest.

In section, brachiopod shells can be seen to consist of several layers. In both articulates and inarticulates, the outermost of these is an extremely thin **periostracum**, which is organic and is not preserved in fossils. Below this, articulate shells consist of **primary** and **secondary** shell material, both calcitic but of different crystal structure (Fig. 5.3 ii).

The primary layer is thin and made up of very finely crystalline calcite, whereas the secondary layer (sometimes called the **fibrous** layer) is built up of closely packed rods or laminae. They lie at an angle to the shell surface, so the primary layer cuts across them rather in the manner of an angular unconformity. Secondary shell is produced throughout the life of the animal, and is responsible for the thickening which occurs towards the umbones.

How is this complex structure produced? The answer lies in the growth of the mantle itself. At the anterior margin, around the edges of the shell, there is a mantle fold. Here we find the generative zone where new mantle cells are formed, thereafter to move outwards towards the leading edge of the mantle. As soon as they form they begin to secrete the periostracal layer. Then as they reach the outermost point, they change over to calcite secretion.

This forms the primary shell. The cells are now fixed in position, and they continue to form primary shell until it has attained the desired thickness. At this point they switch functions for the last time. Each cell becomes responsible for the secretion of a single fibre of calcite, and it will continue making this contribution to the secondary shell until death.

Shell secretion, while being a continuous process, is subject to certain minor fluctuations in rate. It is these which give rise to the growth lines visible on the exterior. These are always present, but may be more obvious in certain species.

An important feature of some articulate shells are certain perforations which penetrate from the interior almost to the periostracum (Fig. 5.3 iii). They are called **punctae** and in life they contain extensions of the mantle. The heads of punctae connect with the periostracum by fine threads passing through the primary shell. They are often regularly spaced over the shell surface, and it has been suggested that their function was sensory, or possibly to deter predatory borers in some way.

They are very haphazard in their taxonomic distribution within the phylum, and like the spiral brachidia mentioned earlier it would appear that they evolved separately many times. Many extant (still-living) brachiopods are punctate, but even so their purpose remains far from clear.

The inarticulate shells are of variable composition, some being chitinophosphatic, some calcareous. But even within these compositional groups the microstructure may not be consistent. In forms like *Discinisca*, for example, the chitin and phosphate are mixed together, whereas *Lingula* has alternating layers of each. In the calcareous types, the calcite fibres are irregularly arranged and have no definite pattern. Relatively little is known about the formation of inarticulate shell material.

5.4 Design mechanics of articulate hinges

How do brachiopods open and close their shells? This apparently simple problem gave rise to very many mechanical solutions in the history of the phylum, some of them more satisfactory than others. First, let us take a look at the first component of the system, the hinge.

When a structure like a lid or a door hinges, it rotates about an axis which we call a **hinge axis**. This may coincide, all along its length, with an actual, straight-line hinge mechanism — a piano lid is a good example of such a case. Other hinged structures, such as swivel windows (which do not hinge along any of their edges) pivot instead about two points set some distance in along two opposing sides — points through which the hinge axis passes.

Both these types of hinge are seen in brachiopods. The older system involves the simple, straight-line hinge as is seen, for example, in *Spirifer* (Fig. 5.13 i). Such hinges are called **strophic** hinges, and being intrinsically strong those forms which have them do not have any need of strong teeth.

The **non-strophic** hinge (Fig. 5.6 i) has no hinge line, the axis passing through two points on either side of the mid-line of the shells. Anterior to this axis, the valves diverge from each other when the shell opens, while posterior to it the brachial valve moves either more deeply within the pedicle valve or more closely to it. In such shells dentition is well developed. The teeth firmly locate the valves and do not allow them to gape more widely than the maximum employed in life. Thus they do not disarticulate upon death, and unlike many strophic shells are nearly always found intact. By far the greater number of living brachiopods are non-strophic.

To close the shell, the adductor muscles must act upon points which lie anteriorly to the hinge axis — just as an elastic band wound around the blades of a pair of scissors draws them together. There are actually two types of adductor muscle present in any brachiopod. One is the **catch adductor**, which can hold the valves tightly closed for long periods, and the other is the **quick adductor**, which reacts very quickly to impulses such as those from the sensitive mantle edge, and can snap the shell shut with great speed.

Opening the shell is more of a problem, because the diductor has to attach at one end to a point which lies behind and/or below the hinge axis. This is because, as we have said, muscles can only function by contraction — there is no such thing as an expanding muscle. All skeletal hinges function by the exertion of forces of contraction on opposing sides of a fulcrum.

Clearly, the muscle could pass outside the shell with the pedicle and attach to the region below the beak of the brachial valve. This simple solution (Fig. 5.6 ii, a), using the delthyrium and notothyrium to allow the muscle to pass outside the shell, obviously impairs the shell's protective efficiency.

To give the muscles greater protection, other brachiopods covered them with shelly plates which were projected backwards so as to enclose them. These posterior projections of the valve edges could not cover the muscles completely when the shell was shut, or else they would have prevented it from ever opening. Therefore, when the valves were closed a small gap between these delthyrial covers would have been created — a slight impairment of protection, but an advance on the system described above (Fig. 5.6 ii, c).

As an alternative to these mechanisms, some brachiopods allow the umbo of the brachial valve to become incurved by slowing down growth along its posterior edge. This tucked-in umbo can then serve as an attachment site for the diductor which will be inside the protection afforded by the pedicle valve (Fig. 5.6 ii, b). This system is most common among living brachiopods.

The most sophisticated solution was effected by attaching the diductors to projections from the valve surfaces which, by their extreme prolongation, could alter the line of action of the muscle. This is an additional function for muscle platforms, whose purpose in highly convex shells has already been touched on. This method eliminated the need for delthyrial covers and kept all the living tissue securely within the shell (Fig. 5.6 ii, d). Once again, this is a feature which we see evolved separately in many unrelated brachiopod groups.

5.5 Mode of life

All brachiopods are benthic marine animals, and most are unable to tolerate the influence of fresh water. Creatures which have a limited range of tolerance to salinity fluctuation are referred to as being **stenohaline**. *Lingula*, on the other hand, is an exception in that it can withstand brackish coastal waters. It has a wide salinity tolerance, and is called **euryhaline**.

A major consideration in a brachiopod's mode of life is its type of attachment. In fossil forms we therefore must ask ourselves the question 'Was it attached, and if so, by what means?' Many brachiopods were free-living, but to what extent could they move? And did any live partially or completely buried within the sediment?

Many articulate brachiopods live attached by a pedicle. This is usually a short cord of connective tissue clothed in tough cuticle. It usually leads to a rock or to another shell, but some are known which are root-like and are able to anchor the animal in soft sediment. The brachiopod may be able to swing round by using its adjustor muscles so as to face different ways according to the current, but in some species the pedicle becomes atrophied and can only function as a tether. In some fossil forms it probably withered away entirely, so that the mature beast was free-lying.

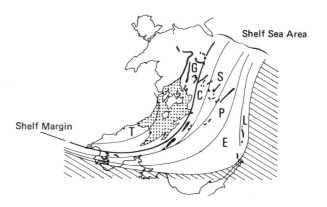

FIGURE 5.7. Llandovery palaeogeography and brachiopod communities in Wales. Cross-hatching, land area. T = turbidites, G = graptolitic facies. C = *Clorinda* community, S = *Stricklandia* community, P = *Pentamerus* community, E = *Eocoelia* community, L = *Lingula* community: in order of decreasing water depth. Dot-shaded and black areas give outcrop coverage of Llandovery rocks (L. Silurian). Redrawn from Ziegler, Cocks and Bambach, 1968.

Free-lying forms (Fig. 5.8) were generally of a shape which ensured their stability. Rather in the way a self-righting toy always comes to rest in the correct orientation, their shape contrived to be unstable in all positions except the one desired. A few free-lying forms may have been motile. The design of the shell of *Chonetes* (Figs. 5.8, 5.12) suggests that it could jet-propel itself backwards by snapping its shells closed and expelling the water from its mantle cavity.

Both articulate and inarticulate brachiopods are known to have taken to cementation as a mode of fixing themselves. In this life style there is a strong chemical bond between the fixed valve and its substrate. Cemented brachiopods were unable to move once they had settled, nor could they change their orientation like the barnacles can, for instance. It is therefore likely that their larvae were programmed to select sites where there was consistent water flow, and to begin growth in precisely the correct alignment relative to it. Cemented brachiopods tend to be found in 'cryptic' habitats, such as beneath boulders, coral colonies and in reef cavities.

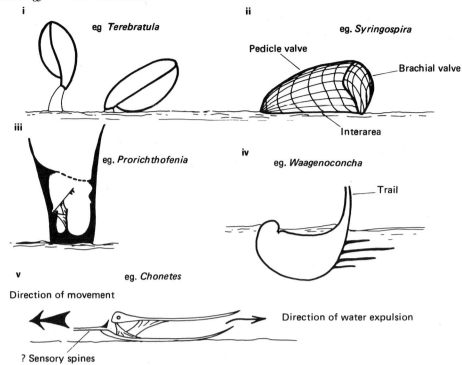

FIGURE 5.8. Brachiopod modes of life. (i) Pedicle attachment. (ii) Free-lying. Expansion of the inter-area lends stability. (iii) Cementation. Gross enlargement of pedicle valve to form a cone within which the brachial valve flaps up and down creating currents. (iv) Semi-infaunal. Concavo-convex form, stabilizing spines. (v) Free-lying, motile. (Mostly after Rudwick)

No brachiopods ever burrowed very deeply into sediment because they needed to maintain contact with the water and were not able to form tubes by fusing their mantle edges together. *Lingula*, perhaps, comes closest to achieving this, in the way that its setae are able to form short siphons (see above).

There is a difference, however, between active burrowers like *Lingula* and forms which merely allowed themselves to become half-buried. Such types were typically concavo-convex in shape, and lived with their narrow, slit-like apertures projecting up through the sediment (Fig. 5.8). They may have had spines to help with anchorage, and are typified by the many strophomenides (Figs. 5.4 iii, 5.11) which adopted this mode of life.

All filter feeders need to ensure that as little water as possible is recycled between the inhalent and exhalent currents. In brachiopods this has led to a very common feature, namely the folding of the commissure into (usually) a median upfold and a pair of flanking downfolds (Fig. 5.4 iv). This pattern of folding separates the two inhalent streams from the exhaust current rising medially. It is a feature which may be most significant to brachiopods living in very quiet water conditions where recirculation would be an even greater danger.

Another common feature of the commissure, seen especially in the rhynchonellides (Fig. 5.16), is the zig-zag. This is often combined with the larger scale of folding just described and is principally a protective measure to prevent particles which might damage the lophophore from entering the shell. By zig-zagging the commissure, its length is much increased, and so a relatively large amount of water can flow even with quite a small angle of gape. This has the additional advantage of keeping the mantle edges close together, ensuring a closer monitoring of the inflowing streams.

An important facet of brachiopod ecology has been revealed by the study of benthic assemblages (Fig. 5.7). These are groups of species which tend to occur together because they favour the same environmental conditions. It has been found that these important environmental factors (temperature, turbulence, etc.) are linked principally to water depth, and bands of depth-related communities may be traced, running parallel to ancient coastlines.

The first of these studies, and still the classic of its kind, was carried out on rocks of the L. Silurian in Wales. Since it was published in the mid-sixties, similar depth zones have been identified in other areas and for other systems. This pioneering work on brachiopods has shown us that depth, as expressed by important physical parameters, has exerted a strong influence upon benthic life throughout all geological time.

Brachiopods 55

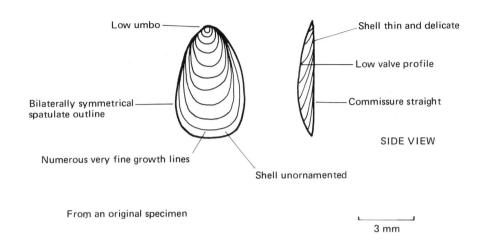

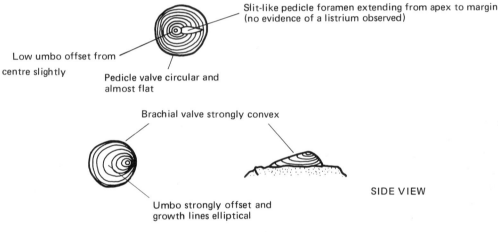

FIGURE 5.9. *Lingula* sp. and *Orbiculoidea* sp.

5.6 Classification

Until quite recently the classification of this important phylum was extremely unsatisfactory. It relied upon groupings which were defined on very few characteristics of form — characteristics which, it was believed, deserved particular weighting. The result was a most unnatural and artificial taxonomy.

The present system, laid out in Table 5.2, is based on that employed in the *Treatise*. Rather than by making *a priori* (beforehand) decisions about what should constitute a major unit, this system has been created by the steady grouping together of species and genera with similarities and inferred relationships. The orders into which the two brachiopod classes are divided are, as a result, rather hard to define by reference to a few 'key' characteristics. Nevertheless, you should find no difficulty in recognizing the members of any of these orders after studying a few typical representatives of each (Table 5.2, Figs. 5.9–5.16).

The Inarticulata are divided into five orders, of which two are geologically significant. These are the O. Lingulida and the O. Acrotretida (Fig. 5.9). Lingulides are oval or spade-shaped, with two gently convex valves of subequal size between which a

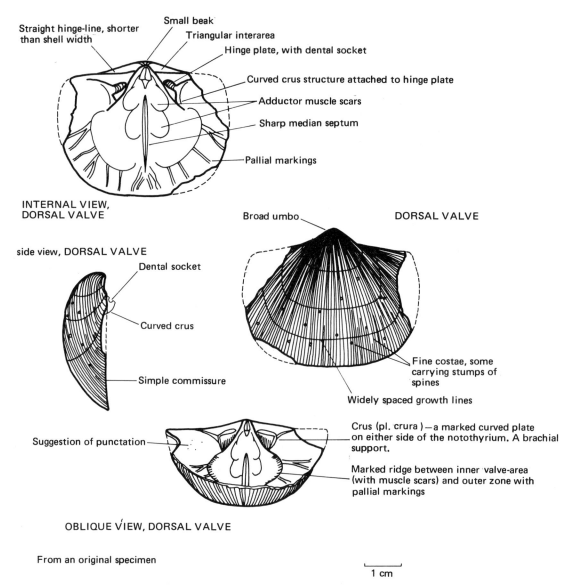

FIGURE 5.10. *Schizophoria* sp.

pedicle may emerge via a shared opening. In others it may be absent. The acrotretides are circular in general outline and the valves are of unequal size and/or convexity. The pedicle foramen penetrates the pedicle valve only, but may be absent. Many are cemented forms, like *Crania*.

The Articulata are divided into six orders. The Orthida (Fig. 5.10) were probably ancestral to all others and have gently convex valves with fine radial ribs and broad, strophic hinges. The Pentamerida were very convex and employed distinctive muscle platforms in both valves (Fig. 5.5). Possibly derived from them were the Rhynchonellida (Fig. 5.16) whose shells are also strongly convex. They are non-strophic and have plicate, zig-zag commissures.

Like the rhynchonellides, the Spiriferida had tri-lobed shells, and they are easily recognizable by virtue of their much-extended strophic hinge lines (Figs. 5.13, 5.15). Other spiriferides (belonging to the Suborder Atrypidina) are non-strophic, however, and they bear fine radial ribs — like the famous and abundant Palaeozoic genus *Atrypa* (Fig. 5.15).

Possibly derived from the Atrypidina were the Terebratulida (Fig. 5.14), the most abundant of the modern orders. They have smooth, biconvex shells, with a pedicle foramen.

Lastly, and most fascinating of all, perhaps, is the O. Strophomenida (Figs. 5.11, 5.12). These distinctive brachiopods, concavo-convex in most cases and often infaunal in habit, are particularly well known

Productus sp.   L. Carb.   Locality – Llangollen, Wales   (O. STROPHOMENIDA)

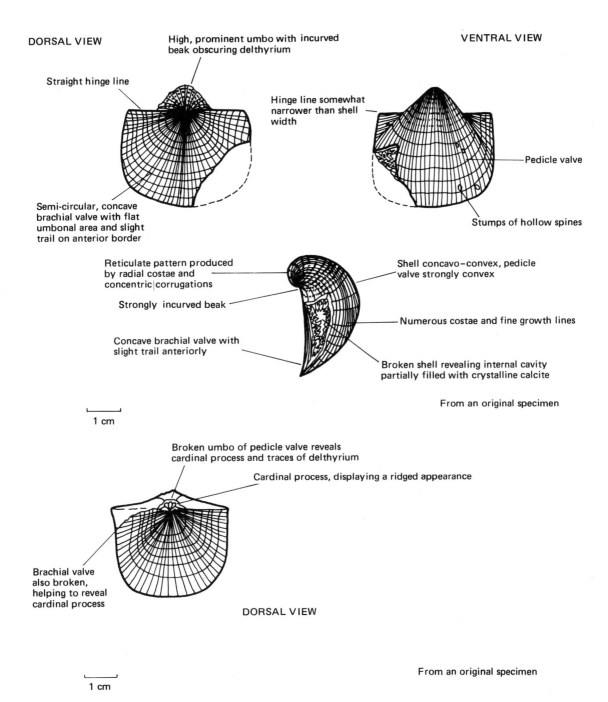

FIGURE 5.11. *Productus* sp.

from the Carboniferous Limestone. Theirs is a large and diverse order, characterized by strophic hinges, spiny exteriors and an extraordinary adaptability. The first articulate brachiopod to adopt a cemented mode of attachment — the Silurian species *Liljevallia gotlandica* — was a strophomenide.

Examine closely the figures which illustrate typical examples of these orders. Make copies of them to familiarize yourself with their morphology. Try to place as many of your own specimens as you can into their correct orders, and make scaled, annotated drawings of them like those in Figs. 5.9–5.16. You should label as much morphological detail as possible.

You might also find it useful (and not only with regard to the brachiopods) to read through the British Museum guides to British fossils (see Suggested Further Reading, p. 164) where you will find a wide range of common species beautifully presented.

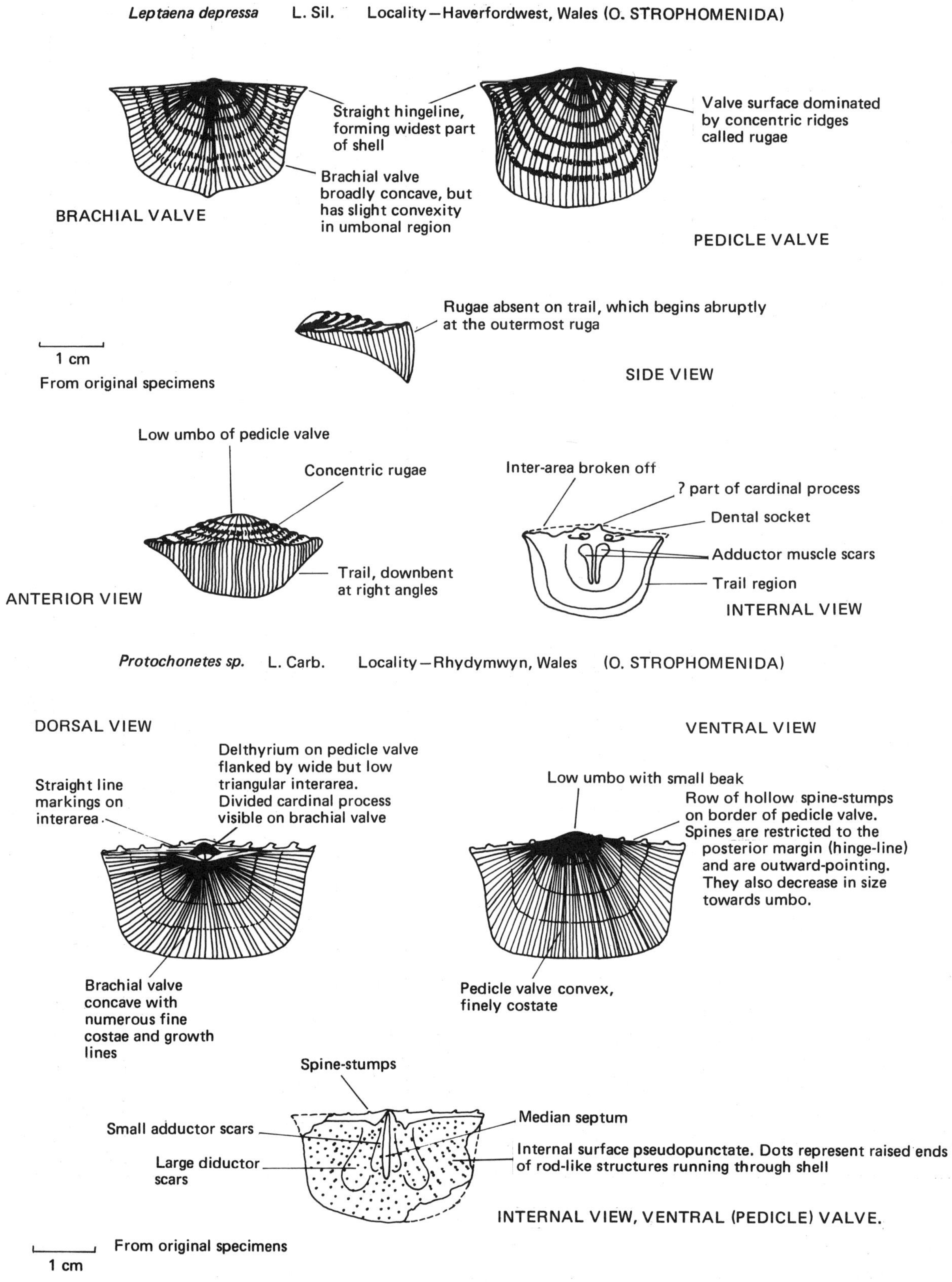

FIGURE 5.12. *Leptaena depressa* and *Protochonetes* sp.

TABLE 5.2. *The Classification of the Brachiopoda (based on the Treatise)*

Class	Order	Important sub-order	Range	Examples
Inarticulata Lacking teeth & sockets. Shell held together by muscles. Most shells chitino-phosphatic	**Lingulida**—biconvex, valves elliptical without specialised foramen		Camb.–Rec.	*Lingula, Lingulella, Lingulopsis*
	Acrotretida—Circular or subcircular. Valves conical with high central umbo. Pedicle foramen opening through pedicle valve		L. Camb.–Rec.	*Crania, Orbiculoidea*
Articulata Valves of shell located with teeth & sockets. Shells calcitic	**Orthida**—Mostly impunctate, biconvex with a straight hinge-line and pronounced cardinal areas. Few punctate. Foramen affects both valves: deltidium usually lacking		L. Camb.–U. Perm.	*Heterorthis, Salopina, Schizophoria*
	Strophomenida—Mostly pseudopunctate—i.e., shell structure characterized by puncta-like structures which, however, did not contain mantle caecae. Shells concavo–convex with straight hinge-lines. Most forms lose pedicle-muscle function at adulthood	Strophomenidina—features as for the order	Ord.–Trias.	*Leptaena, Rafinesquina, Sauerbyella Strophomena*
		Chonetidina—row of spines along hinge-line of pedicle valve. Free-lying, possibly motile.	L. Sil.–L. Jur.	*Protochonetes, Rugosochonetes*
		Productidina—pedicle valve spinose. No pedicle muscle in mature forms. Some species lost cardinalia, teeth and sockets.	L. Dev.–U. Perm.	*Productus, Gigantoproductus, Richthofenia*
	Pentamerida—Biconvex, impunctate with large muscle platforms		M. Camb.–U. Dev.	*Gypidula Conchidium Pentamenis*
	Rhynchonellida—impunctate biconvex shells with internal crura to support lophophore. Radial ribs a common feature. Foramen restricted by deltidia		M. Ord.–Rec.	*Camarotoechia Goniothynchia Pugnax Tetrathynchia*
	Spiriferida—punctate and impunctate biconvex shells with spiralia to support lophophore. Foramen usually restricted by a single deltidial plate or pair of plates	Atrypidina—Impunctate, with non-strophic hinge and characteristic finely costate shells.	M. Ord.–U. Dev.	*Atrypa, Atrypella, Atrypina*
		Athyrididina—Impunctate, similar to the above, but spiralia point laterally rather than dorso-ventrally and are linked by a 'jugum'.	U. Ord.–L. Jur	*Meristina, Pentagonia*
		Spiriferidina—Punctate or impunctate. Spiralia directed laterally but lack a jugum. Hinge line strophic, shells ribbed and with central fold.	L. Sil.–L. Jur	*Cyrtia, Spirifer, Spiriferina*
	Terebratulida—biconvex punctate shells with pair of cruralia and loop to support lophophore. Foramen usually at beak of pedicle valve and restricted by deltidial plate or pair of plates	Terebratulidina—Short loop extending from the crura.	L. Dev.–Rec.	*Carneithyris, Dielasma, Terebratula, Lobothyris, Terebratulina*
		Terebratellidina—Long loop—often recurved—extends from the crura and the dorsal medium septum.	L. Dev.–Rec.	*Digonella*

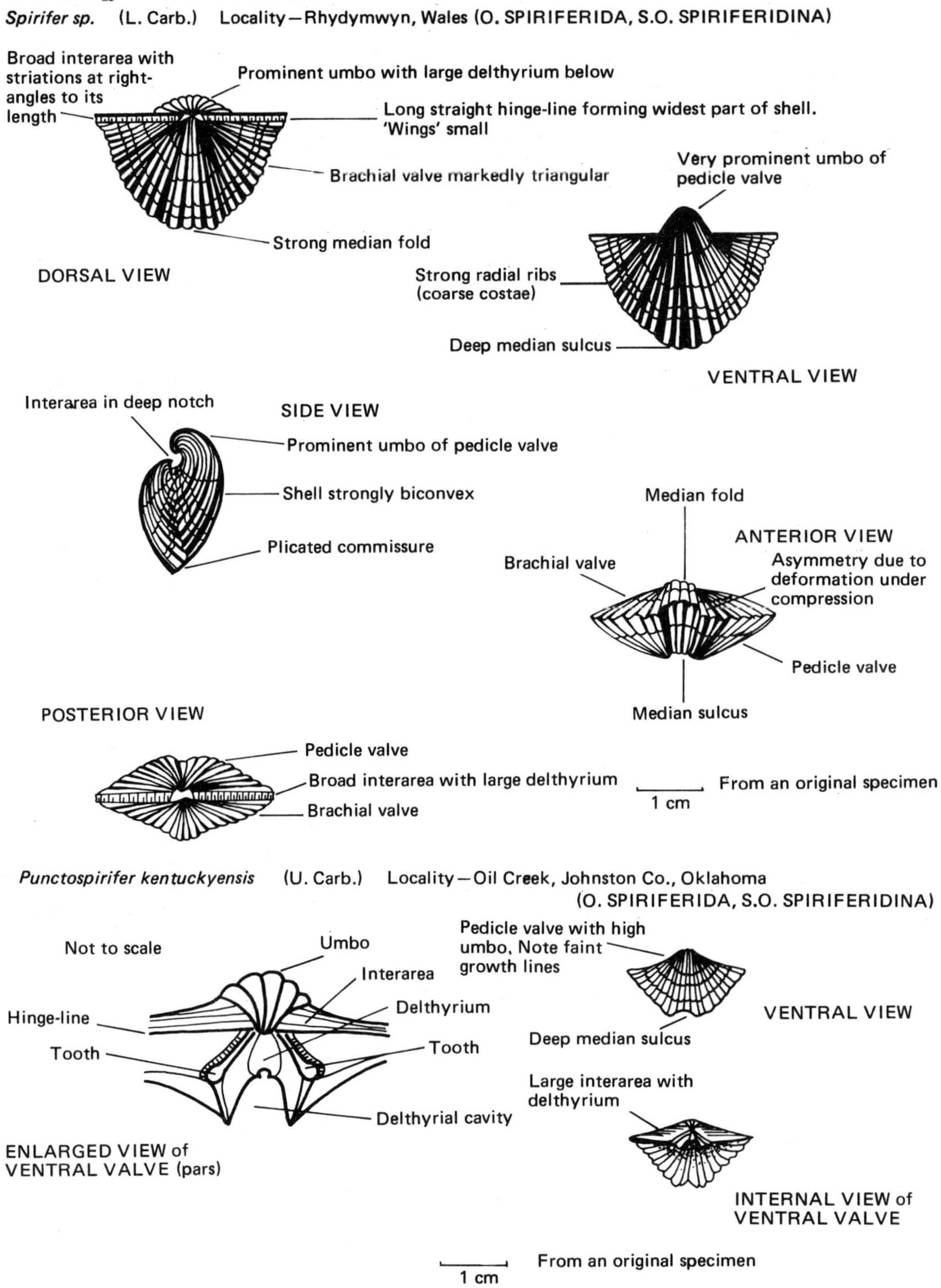

FIGURE 5.13. *Spirifer* sp. and *Punctospirifer kentuckyensis*.

5.7. Geological history and stratigraphic value

The brachiopods, like most fossil groups, appear suddenly in the Lower Cambrian, and there are no Precambrian fossils which might be described as precursors. (Early reports of 'protobrachiopods' have now been disproved, upon the discovery that these supposed fossils were actually sedimentary structures!)

All orders of inarticulates are present in Cambrian rocks, and all but the two which survive at the present day became extinct before the end of the Silurian. The articulates, represented by the orthides in the first instance, only became truly dominant in the early Ordovician, when they underwent their wide-ranging diversification (see above).

By the end of the Ordovician, all orders of articulates except the Terebratulida had appeared.

Ornithella sp. U. Jur. Locality—Abbotsbury, Dorset (O. TEREBRATULIDA)

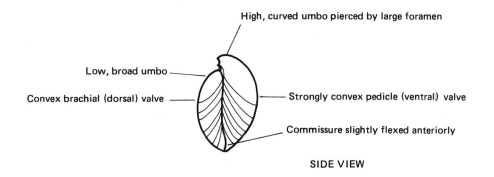

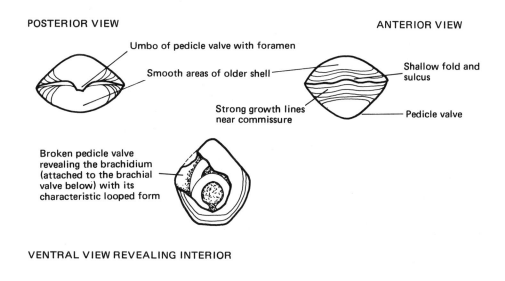

FIGURE 5.14. *Ornithella* sp.

These came in only in the late Silurian. There was a minor decline in the Devonian, involving some extinctions (the pentamerides, for instance), but further radiation took place in the Carboniferous, when the productides (a suborder of the Strophomenida) began to develop some very unusual designs to exploit some very novel habitats. Many lived quasi-infaunally, but in the Permian the richthofeniids (Fig. 5.8 iii) became cementing, reef-building organisms. Their pedicle valve was enormously enlarged into an inverted cone, closed by an inset brachial valve which probably flapped up and down to create feeding currents.

With forms such as these the brachiopods were at the peak of their adaptability when the mass extinctions at the close of the Permian all but wiped them from the face of the earth. Although they are present (and occasionally very abundant) in the Mesozoic

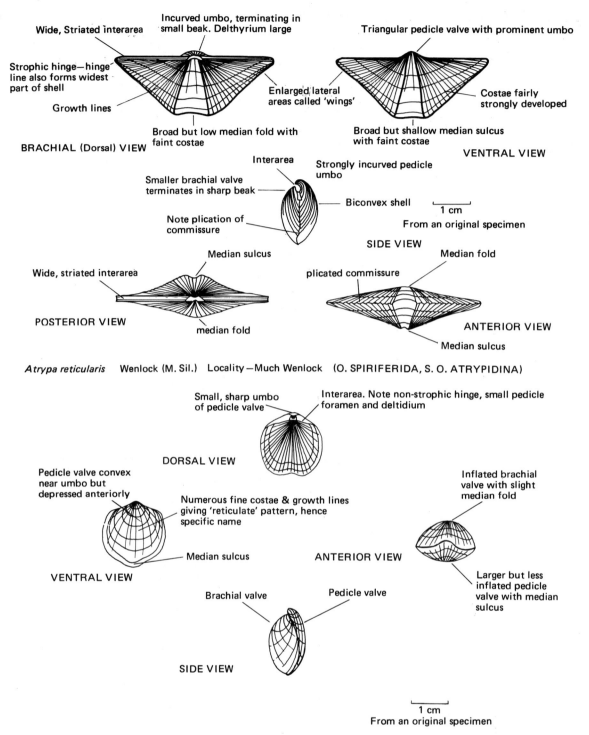

FIGURE 5.15. *Mucrospirifer thedfordensis* and *Atrypa reticularis*.

rocks, the diversity of orders was never re-established, and the phylum seems to have been in decline right from the end of the Palaeozoic.

The reason usually cited for this is the immense diversification of the bivalved molluscs (Ch. 6), which are presumed to have displaced them ecologically. But this cannot be the whole truth, because the bivalves owe their success mainly to their exploitation of the deep-burrowing habit, one into which the brachiopods never truly entered.

Brachiopods, together with trilobites, have been used as guide fossils in Palaeozoic shelf sequences where graptolites are rarely found. Ordovician and Silurian graptolite zones are well coordinated with corresponding assemblages of shelf fossils, and indeed in some cases the degree of refinement which they

Goniorhynchia boueti (M. Jur.) Locality—Boueti Beds, Herbury, Dorset (O. RHYNCHONELLIDA)

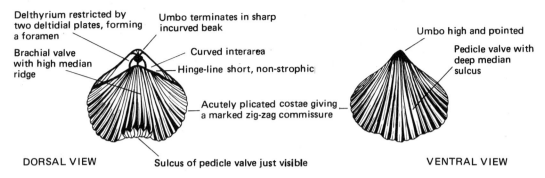

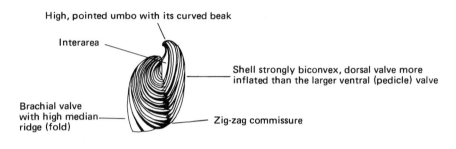

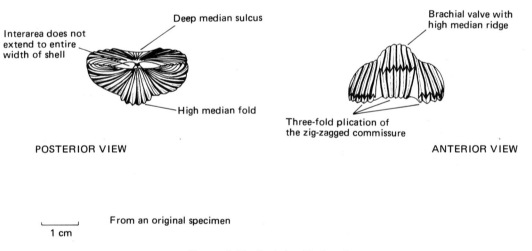

1 cm   From an original specimen

FIGURE 5.16. *Goniorhynchia boueti*.

afford can be even greater than that of their deep-water counterparts. The chief problem with brachiopods, as with most benthic animals, is their provinciality and susceptibility to changes of facies.

Assemblages, rather than individual species, tend to be used, though certain species are useful markers, at least over limited areas — such as within Britain. *Conchidium knighti*, for example, is a reliable marker for the Ludlow (U. Silurian) and *Stringocephalus* faithfully marks the Middle Devonian.

In the Mesozoic, the stratigraphic use of brachiopods is further hampered by the longevity of the species. One notable exception to this is the rhynchonellide *Goniorhynchia boueti*, which is restricted to a band about 45 cm thick in the Great Oolite Series of the English Jurassic.

6. Molluscs

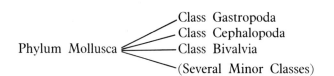

6.1 Introduction

The idea of nature's diversity is so familiar to us that it is especially exciting when science reveals her underlying unity. Such is certainly the case with the molluscs, which comprise organisms so apparently different from each other that one would hardly guess their common kinship.

Snails, slugs, cockles, mussels, limpets, cuttlefish, octopuses and a host of other forms familiar and unfamiliar, living and fossil, are all united in this great phylum, which is the second most diverse after the Arthropoda. 80,000 living and 35,000 fossil species cover a huge range of habitat from deep seas to mountain tops. There are browsers and filter feeders and predators, ranging in size from minute plankton to the largest invertebrates known. The giant clam *Tridacna* may weigh a quarter of a ton; fossil nautiloids are known which are up to 10 m in length, and the giant squid, while captured at 16 m maximum, is thought from sucker marks seen on whales to grow to 40 m!

Yet this stupendous variety of form, size and mode of life has been developed upon an anatomical blueprint which (though altered here and there) remains recognizably constant throughout. This is the special fascination of the Phylum Mollusca.

6.2 General molluscan anatomy

In this section we shall not be dealing with the anatomy of any *known* mollusc. Rather, we shall examine the ground plan of all the various anatomies which are now seen in the different classes, a standard model from which all subsequent ones have been refined. Although no living mollusc is known to possess such a design, it may well be that at some time in the Precambrian there existed an early mollusc not unlike this.

Protecting our hypothetical mollusc (Fig. 6.1) is a shell of calcium carbonate. It is lined inside with a secretory coat which also encloses the body. This is called the **mantle** (or **pallium**). The space which it and the shell together enclose is not entirely filled by the body, and there is an empty volume posteriorly called the **mantle cavity** into which the gills hang and the anus opens. Water is drawn in on one side, passes over the gills and, on its way out, collects the waste material and carries it away.

Not all molluscs display them, but it is an ability of this phylum to fuse the mantle edges together and extend them into long tubes called **siphons**. These are used to carry inhalent or exhalent currents and have contributed greatly to the success of the phylum in marine habitats. We shall see these structures being of great importance to both gastropods and bivalves.

In front of and above this mantle cavity lies the visceral mass, which contains a simple heart, the kidneys and the gut. The gut opens anteriorly at the mouth, in which there is a rasp-like tongue bearing sharp, inward-pointing teeth. It is called the **radula**, and it is used for grazing food (typically algal films) off the substrate. Digestion takes place in an out-pouching of the gut called the **digestive gland**, and waste is expelled as discrete **pellets** via the anus. These minute faeces can be very important sediment-builders, especially in some limestones.

The 'brain' consists of a nerve ring surrounding the oesophagus, and two pairs of nerves arise from it. One pair serves the gut and another serves the large, muscular **foot**, upon which the animal rests and moves. The head is situated anteriorly upon this organ, and as well as the mouth it also bears primitive eyes. Other sense organs include simple balancing organs called **statocysts** (in the foot) and a pair of **osphradia**, which may be sensitive to water chemistry and monitor the sediment content of the inhalent current.

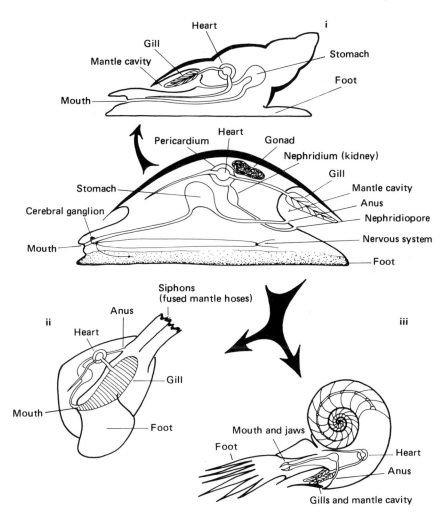

FIGURE 6.1. The common features of the molluscan body plan, as they may have derived from a hypothetical common ancestor displaying basic mollusc structures in an 'unspecialized' form (centre). (i) The gastropod. Note the torsion of the shell and upper part of the body to bring the mantle cavity to the anterior. (ii) The bivalve. Note reduction of head and the mouth attached to the gill as part of the filter-feeding mode of life, and the fusion of the mantle to form siphons. (iii) The Cephalopod. Note the subdivision of the foot to form tentacles and the coiled, chambered shell. Mouth is provided with heavy jaws for predation.

Reproduction in such an 'ancestral' form would probably have involved the release of eggs and sperm into the water from separately sexed individuals. After the external fertilization a brief larval life would have led to metamorphosis and eventual adult development.

6.3 A diversity of molluscs

Molluscs probably evolved from segmented ancestors. Segmentation is the repetition of a pattern of elements along the length of the body, each containing a replication of the preceding segment's nerves, blood vessels, appendages, etc. It is an evolutionarily primitive condition and can be seen very clearly in annelid worms, though we have already encountered it in trilobites. In the trilobite thorax we noted that the segments are easily observed, but that they have been variously suppressed in the cephalon and pygidium (section 3.3).

In many molluscan classes segmentation is so effectively suppressed that it no longer appears. However, molluscs are known that still possess bilaterally symmetrical bodies, where certain anatomical elements (such as the muscles which hold the animal in its shell) are repeated in pairs. The ancestral mollusc would, in all probability, have preserved some degree of this segmentation.

The many alterations to the body plan which we shall examine are mostly the result of changes in the dominant mode of life adopted by the group concerned. For example, the bivalves have no need of a head since they live mostly as filter feeders buried in sediment. Their foot is modified for burrowing and so their locomotory abilities are limited, while their gills have been modified for food-gathering and the mantle cavity enlarged. To maintain contact with the water they have developed siphons, which they use like snorkels.

The gastropods, which show least deviation from the basic blueprint, seem to have remained so unchanged because they have preserved grazing as their typical mode of life. They thus still have a radula, an anterior mouth and a fairly well-developed head. Many cephalopods, on the other hand, have become active hunters needing acute senses and high speed. They have evolved large, efficient eyes and converted the mantle into a jet propulsion system which can eject water quickly through a nozzle called the **hyponome**. Thus released from the role of locomotion, the foot has been divided into **tentacles** for grasping prey.

We shall look at these modifications again as this chapter progresses. The separate classes of the Mollusca are each as important as many another whole phylum, though their biological unity demands a common treatment. Such a treatment can also be extended to the description of their shells, which is, of course, of great importance to the palaeontologist.

6.4 Growth in coiled shells

Many animals, not just molluscs, can increase the size of their shells as they grow and so avoid having to moult their exoskeletons as the arthropods do. The advantage this conveys in terms of protection is obvious.

In the brachiopods and molluscs we see a system of growth controlled by the edge of the mantle; growth can only take place around the shell aperture. It should be self-evident that the simplest form of shell which this system could produce would be a cone, as growth proceeds steadily from the post-larval stage to maturity.

If growth were rapid, then the cone would expand greatly in a short distance — as is seen in the limpet, for example. Conversely, if the rate of addition to the aperture were not accompanied by any great increase in diameter, then a long, tubular kind of cone would result. And if the rate of forward growth of the shell were to be slightly greater on one side than on the other, the shell would curl up and assume some sort of curved or spiral form.

Such coiling may take place in one plane (like a Catherine wheel) or it may be a three-dimensional coiling, like that of a helter-skelter (helical coiling). But all these types of shell would be producible from growth at a single, widening aperture.

We can define certain properties of growth which, in varying combinations, can be used to explain all the shell forms which nature produces. They have already been mentioned above; let us now formulate them properly (Fig. 6.2).

1. The rate of increase in size of aperture relative to the speed of shell growth is called the RATE OF EXPANSION OF THE APERTURE.
2. When a Catherine wheel is rolled up, the coiling (being all in one plane) involves an increase in the diameter of the coil by one thickness of the gunpowder tube per revolution. Shells tend to be more complex. Remember that the tube is expanding all the time by a rate defined in (1) above. Also, each coil may become partly embedded in the succeeding coil by a certain amount. We need, therefore, to define precisely how far, per revolution, the aperture travels away from the coiling axis. This is the RATE OF CHANGE IN DISTANCE OF THE APERTURE FROM THE COILING AXIS PER REVOLUTION.
3. We have said that some shells (like that of the snail) coil helically, moving downwards as they spiral. The spiral may be a very tall one, like a helter-skelter, or it may be a very gentle one, producing a shell shaped like an upturned saucer. In other words, the aperture may travel rapidly down the coiling axis, or it may not. Once again we must define the rate of downward movement of the aperture per revolution about the axis. This is called the RATE OF WHORL TRANSLATION.

We can now use these rate controls to define many different shells. The limpet has a very high rate of apertural expansion. So although it grows more rapidly on one side than another (a limpet's apex is rarely in the centre of the outline) one revolution about the axis of coiling is never completed before maturity.

A brachiopod has two valves, and each of these is a coiled shell. The same is true of bivalves. A brachiopod valve, say the pedicle valve of *Terebratula*, expands very quickly so that it reaches its full apertural diameter in a little over one revolution. Since the coiling is in one plane, the resulting shell is bilaterally symmetrical.

A bivalve shell, by contrast, has similar rates of expansion combined with an element of translation which results in the asymmetry (see Fig. 6.2 iv, vii).

In the case of *Helix*, the garden snail, the aperture moves down the coiling axis as the aperture continuously expands through several rotations. Succeeding coils come to lie below their predecessors. It should not be hard to imagine that a low rate of whorl expansion per revolution, combined with little change in distance of aperture from coiling axis, would give rise to a long, pointed shell (Fig. 6.2 ii).

Once you have grasped this system, the variety of molluscan shell form will appear much less bewildering. Of course not all morphological features are accountable in this way; elements like surface ornament, apertural spines and other outgrowths are special features which must not be overlooked. They

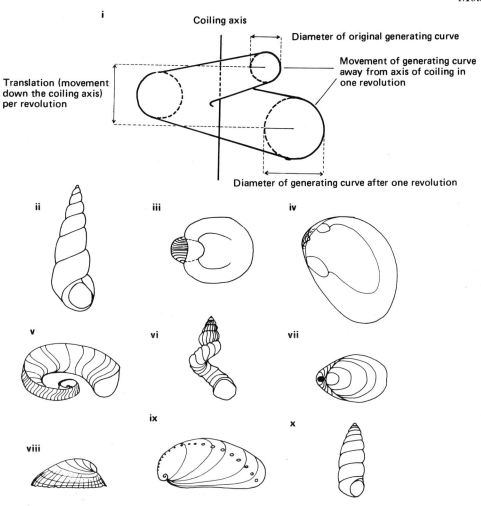

FIGURE 6.2. (i) Schematic representation of a coiled shell to demonstrate the four parameters which define its growth. (Modified from Raup, D.M., 1966.) (ii) Turreted or high-spired gastropod with low rate of whorl expansion and high rate of translation. (iii) Bellerophontacean gastropod showing involute planispiral coiling. Translation zero, whorl expansion higher than rate of movement of generating curve away from axis. Early whorls embedded in later ones as a result. (iv) Bivalve with high rate of whorl expansion, slight translation. (v) Disjunct coiling. Translation zero (planispiral), movement from axis exceeds rate of whorl expansion and whorls lose contact. (vi) Increase of translation at maturity creates disjunct coiling. (vii) Brachiopod showing planispiral coiling with very high rate of whorl expansion. (viii) Limpet-type shell, high expansion rate. (ix) Increase of expansion rate in logarithmic style. (x) Pupiform gastropod showing decreased whorl expansion with maturity.

may have important functions in the species which bear them. Our perception of underlying unity must not divert us from considering the importance of those apparently superficial details which mask it!

I. GASTROPODS

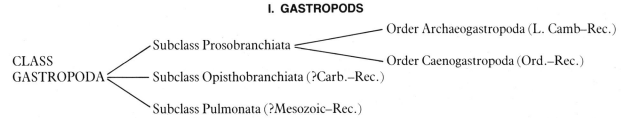

6.5 Soft parts and life-cycle

Gastropods (Fig. 6.3) have a well-developed **head foot** upon which are found the eyes and a mouth containing the radula. The shell is in one piece, and the mantle cavity (except in the pulmonates) contains gills which are used for respiration only. Let us now look at some of the major exceptions.

Firstly, many gastropods have lost their shells altogether — slugs, for example. This means, of course, that they cease to be of palaeontological

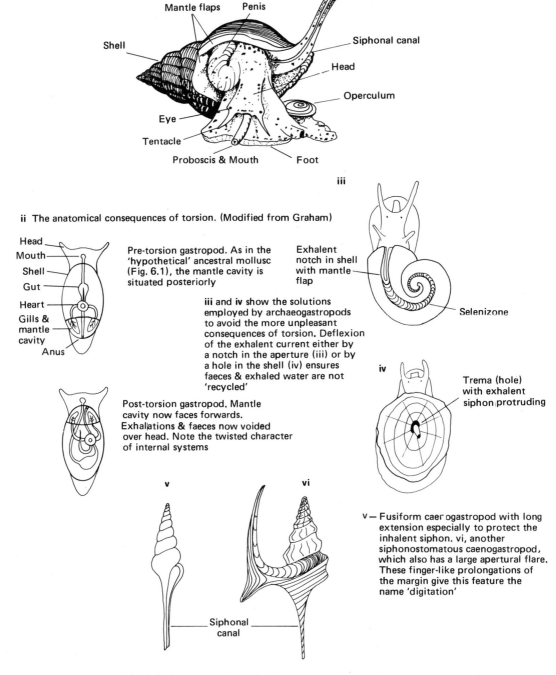

FIGURE 6.3. (i) The whelk *Buccinum undatum* showing relation of soft and hard parts (redrawn from the 'Treatise'). (ii) Cartoons to show the anatomical consequences of torsion. (iii) A generalized archaeogastropod with slit-band (selenizone). (iv) A keyhole limpet cf. *Diodora* to show the alternative 'trema' in archaeogastropods. (v) & (vi) Caenogastropods showing their siphonal canals (compare with i) vastly extended. (vi) also has a digitate margin.

interest. Pulmonate gastropods generally have eliminated their gills and converted the mantle cavity into a lung. This is an adaptation for life on land, but as we shall see, many pulmonates have returned to the water. Another modification seen in pulmonates is the elimination of the larval forms and the introduction of direct development from the egg. An analogous innovation took place in the life cycle of vertebrates when they made the transition from water to dry land, and both point to the importance, for a land animal, of not being tied to the water for purposes of reproduction (see Ch. 9).

The sexuality of gastropods is extremely varied and complex. In the oldest group the sexes are separate and fertilization external. Internal fertilization is seen in other groups, and there may be very elaborate rituals of courtship and strenuous feats of copulation in order to achieve it. By no means all gastropods are single-sexed. Some are hermaphrodite, while others may start off as one sex and become members of the other in later life.

The lack of unity shown in sexuality is contrasted by the fact that all gastropods undergo a process called **torsion** in early development (see section 6.7 and Fig. 6.3 ii). The body mass is twisted, during this procedure, through a full 180° relative to the head foot. The need for this to happen has never been fully explained. Maybe it confers some advantage upon the larva, but pulmonates still display torsion, even though their larval stages have long been suppressed. Adult gastropods have to resort to quite severe remedial modifications in order to overcome the disadvantageous consequences of torsion. Of the many differences in anatomy seen in the various subclasses, a large number have come about as innovations in the struggle against these consequences (see below).

6.6 Shell form

Gastropod shells are made of calcium carbonate, usually in the form of aragonite. Because this is a very soluble mineral you will often find that gastropods have been dissolved out and either left as empty moulds or else infilled by some other substance. Living gastropod shells are coated externally by a thin organic film called the **periostracum**. This is soon lost after burial.

The shell of most gastropods is a conical tube coiled helically about a coiling axis. The shell therefore takes the form of a central pillar, called the **columella**, wreathed in a continuously enlarging spiral chamber (Fig. 6.4). Most are coiled **dextrally** (i.e. if you hold the shell upright with the aperture facing you, the aperture is on the right-hand side), but a few rare cases are **sinistral** (Fig. 6.4 ii).

There is quite a wide range of terms useful in the description of gastropod shells, and they are set out in Fig. 6.4. There are also many typical shell shapes, and these are also shown and given their technical names. You should study and memorize these diagrams. Copy them, and make scaled, annotated drawings of specimens. Remember that these do not need to be fossils! The shells of modern gastropods are plentiful and are ideally suited to the purposes of the student palaeontologist.

In many gastropods the aperture may be sealed by a disc called the **operculum** when the animal retracts into its shell. This lid is attached to the foot (Fig. 6.3) and may be either calcareous or horny. Although rarely found as individual fossils, they may be seen preserved in place, suggesting that the animal retracted for some reason and was killed shortly after — perhaps by a sudden inrush of sediment.

6.7 The anatomical consequences of torsion

There are two major effects which concern us. Firstly, the orientation of the shell relative to the head foot mass is altered, so that in the snail, for example, the shell is sited posteriorly rather than anteriorly above the animal's head. This is doubtless a more convenient arrangement, but the second effect of torsion is decidedly not. The mantle cavity, which originally pointed posteriorly (Fig. 6.3 ii), now opens anteriorly — with the result that the exhalent current, bearing metabolic wastes and faeces, now flows directly on to the animal's head.

Considerable effort has been invested in avoiding this unpleasant consequence. For example, members of the Subclass Opisthobranchiata undergo 'detorsion', reversing the original process. Other solutions, however, are less drastic, and more in the manner of a biological 'patch-up' to make the best of a poor arrangement.

Thus the prosobranchs (whose mantle cavities still face forward) have developed special deflecting devices to separate and divert the inhalent and exhalent streams. We shall now see how this produces recognizable features in the shells of archaeo- and caenogastropods which allow the palaeontologist to identify them. It is not always the case that we are able to apply classifications based on biology, since anatomical distinctions may not be reflected in hard-part morphology. In this case we are lucky.

6.8 The archaeogastropods

Their name means 'ancient gastropods', and they are the parent stock of all subsequent groups. They preserve many of the features of our hypothetical primitive mollusc, including the paired gills which, in more 'advanced' forms are reduced to one. This reduction allows the anus to move further in the exhalent direction and so helps to prevent fouling.

The exhalent current still has to be deflected from the head, however, especially since that is where the inhalent stream enters. It is therefore taken out through a hole in the shell or a deep cleft in the aperture. The presence of a hole (**trema**) or the trace of a slit in the growth lines (**slit band** or **selenizone**) is easily recognizable in fossils (Fig. 6.3 iii, iv).

6.9 The caenogastropods

The success of this group in more recent times has been due to their more efficient gill system which is less sensitive to fouling and has enabled members of the group to occupy niches in soft sediment. Certainly, at the present time, archaeogastropods are relegated to clean waters and mostly rocky substrates.

The caenogastropods have neither selenizones nor tremata. Their exhalent current is deflected away, but to prevent any chance of recirculation with the inhalent stream this is drawn in by way of a long hose called the **inhalent siphon**. Such a tube, if present, is

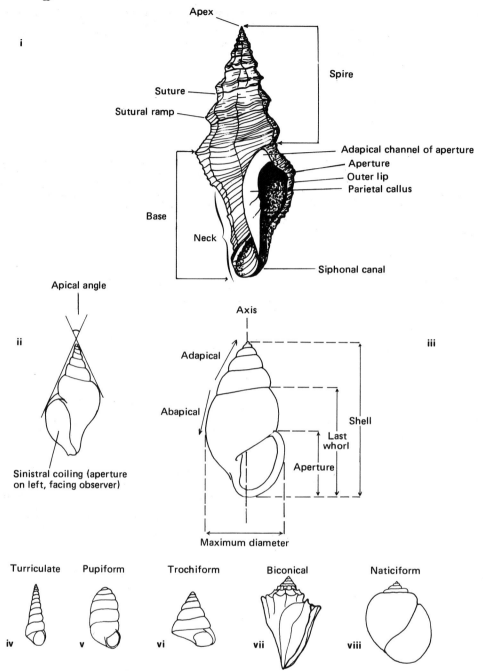

FIGURE 6.4. Morphological features and terms. (i) *Latirus lynchi* showing basic terminology. Note the presence of a siphon, drawing out the aperture abapically. Apertures of this kind are known as 'siphonostomatous'. Entire apertures lacking siphons (iii, below) are called 'holostomatous'. (ii) Sinistrally-coiled shell of 'turbinate' plan, showing the 'apical angle' — a common measurement. (iii) Some more common measurements made in describing gastropods. Note also the positional reference terms 'adapical' and abapical'. (iv)–(viii) Some common gastropod shapes and their names. Note that 'turriculate' forms are commonly also called 'turreted' or 'high-spired'. (Compiled from various sources, mainly the 'Treatise'.)

commonly enclosed in a fold of shell material stretching from the base of the aperture in a long spike. This is the **siphonal canal** (Figs. 6.4 i, 6.5 iv).

The development of the inhalent siphon may have 'preadapted' the caenogastropods for other modes of life. Some burrowing forms employ it as a snorkel, while others, carnivorous in habit, tend to use it as a probing, sensory organ — rather in the way an elephant employs its trunk.

Some rare tropical caenogastropods have become land-dwellers, but most available terrestrial niches have been appropriated convincingly by the next group, the pulmonates.

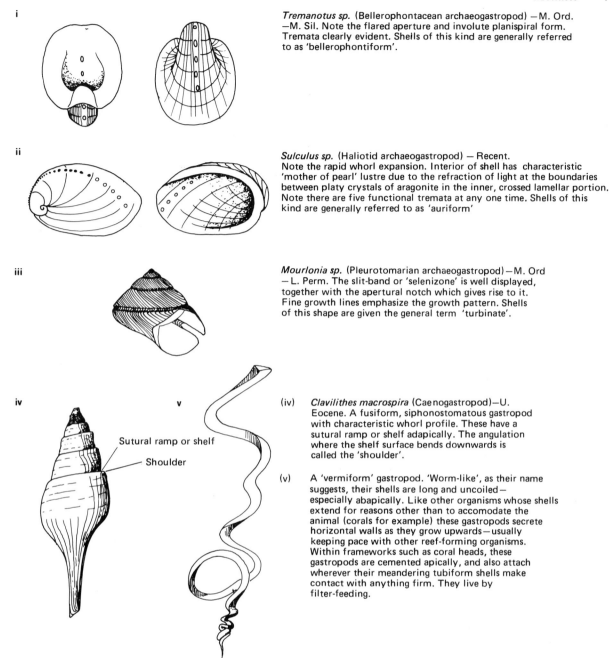

FIGURE 6.5. Some representative gastropods. Not drawn to scale, from the 'Treatise' and various other sources.

6.10 The pulmonates

Pulmonates typically have no gills and instead breathe air directly by the use of a mantle wall highly charged with blood vessels. The meagre 700 or so known fossil species do not do justice to their immense diversity, with over 14,000 living species — including many which have actually returned to the water.

These 'secondarily aquatic' pulmonates either maintain an air-breathing mode of life or else have redeveloped secondary gills (quite different from the gills of their ancestors) by folding the mantle wall so as to increase its absorptive capacity. It is a rule of evolution that, once completely evolved out as gills were in pulmonates, complex structures cannot make comebacks by any kind of 'regression' or 'backward evolution'. They have to be remade from scratch. This rule is borne out by the way in which pulmonates have, so to speak, 're-invented' the gill.

Their return to the water (and they are very numerous in fresh water) has improved pulmonates' chances of fossilization. Nevertheless, recent land snails are extensively used in archaeology as what are termed 'subfossils'. They help to determine environmental conditions in the more recent past (see below).

6.11 Mode of life

No gastropods appear to live longer than about a decade, and for that time the vast majority live by grazing on algae and higher plants in shallow water or in moist conditions on land.

Carnivorous habits are seen among pulmonates, but only rarely; it is the prosobranchs which are the most effective hunters. Certain species attach to bivalve shells and, using the radula (often in conjunction with chemical secretions), are able to bore a small, bevelled hole to enable insertion of a specially adapted proboscis. Through this, the soft parts of the bivalve are slowly ingested. Fossil bivalves bearing these distinctive marks of predation may be found in rocks as old as the Devonian. Other gastropods, preying on bivalves, may prise the valves apart using the muscular foot and employing their own apertural margin as a wedge. Some rare carnivores, such as the conch shell *Conus*, may deliver powerful stings — sufficiently powerful in this case to be dangerous to humans. Many divers in tropical areas have drowned as a result of paralysis brought on by such stings.

Filter feeding is a mode of life usually adopted by fixed forms. We have seen how brachiopods live by this method, and in certain fairly unusual cemented or burrow-dwelling gastropods, a similar system is developed. The American Slipper Limpet *Crepidula fornicata* lives fixed more or less permanently in dense colonies, and has very much lengthened gill filaments to entrap plankton. The high-spired, turriculate (Fig. 6.4 iv) gastropod *Turritella communis* lives buried in mud, its pointed shell lying at about 10° to the sea bed with its aperture protruding. It has somewhat reduced salivary glands and radula, and appears to live entirely by filtration.

Certain aberrant, uncoiled gastropods (Fig. 6.5 v) can cement in early growth and grow rapidly upwards in a loosely coiling manner, attaching to anything which might give them support along the way. They commonly live inside branching coral colonies in tropical regions, and like many organisms which indulge in such competitive upward growth, their shells can become too big for their bodies. They are therefore very unusual among gastropods in that they can wall off portions of their old shell volume with bulkheads called **septa**.

The last mode of life which we shall consider is free-floating. Many gastropods are passive floaters on seaweed or specially constructed rafts, but the pteropods (Subclass Opisthobranchiata) are free-swimmers. They secrete short, straight shells of aragonite which can be important sediment-builders (pteropod ooze) down to about 3500 m. Below this level, aragonite is dissolved and pteropods no longer appear in the sediment.

We have examined the function of certain features in gastropod shells — the selenizone, the siphonal canal, and so on — but what of their overall morphology? Why are there so many identifiable and recurring shapes of shell (Fig. 6.4)? If the animal merely needed to protect its soft parts and to keep them tucked away in a receptacle which was as conveniently designed as possible, then why are not all gastropod shells of roughly the same shape? The fact that the high-spired *Turritella* is a burrower suggests that there must be more functional significance in shell outlines.

This conclusion is supported by the gastropods found in the Silurian rocks of Nova Scotia (Fig. 6.6). These represent cycles of deepening and shallowing of water over millions of years. The study of the sediments identifies three environments; hard-bottom open sea, soft-bottom open sea and soft-bottom lagoon. In each of these environments an assemblage of gastropods having definite shell shapes existed. From one cycle to another, many species became extinct: yet they were consistently replaced by new species which had the same shell morphology. This strongly suggests that certain shapes were very much favoured by certain conditions.

But what shapes and what conditions? To find out we have to compare fossil and living morphological groups. Now that we know, for example, that *Turritella* is a burrower, we may tentatively assume that other turreted gastropods may have been burrowers. But there is much danger in plunging, bare-headed, into this kind of reasoning. We have to consider possible differences between our fossil examples and the modern forms.

Modern *Turritella* is a caenogastropod, so it has gills which are inherently better at dealing with suspended sediment than the gills of archaeogastropods. Even so, experiments have shown that the slightest turbidity in the water will cause it to stop feeding. What, therefore, are we to make of high-spired archaeogastropods? We have noted that caenogastropods have excluded their older-established cousins from modern muddy habitats. But in the past, at a time when they had no competition, could archaeogastropods have occupied the niche in which *Turritella* now lives?

Other deductions are happily less fraught with unknown factors. Size, for example, may be just as 'functional' as shape. Very small gastropods tend today to live in the foliage of seaweed. Of course, seaweed is very rarely preserved, and large numbers of very small gastropods in a sediment may suggest the erstwhile presence of dense, algal groves.

6.12 Evolutionary history and geological value

The first gastropods to be seen in the early geological record are the Bellerophontacea. These are unusual among gastropods in that their shells are planispiral and bilaterally symmetrical (Fig. 6.5 i).

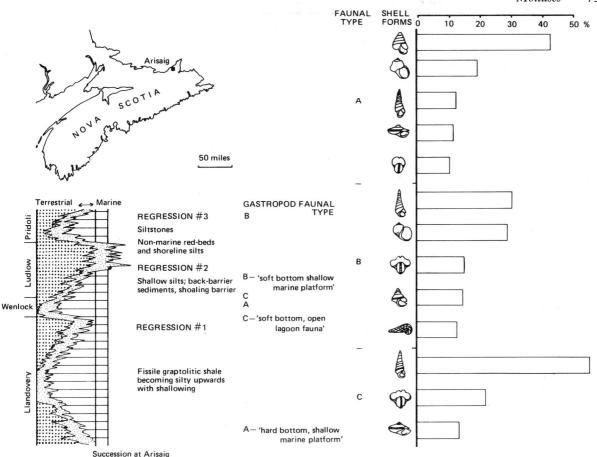

FIGURE 6.6. The Arisaig Group (Silurian), Nova Scotia; stratigraphy, sedimentology and gastropod palaeoecology (redrawn and compiled from Peel, J. S., 1978).

They are problematical for other reasons too, for although some are true archaeogastropods with slit bands, tremata, indented margins, etc., others are known with entire margins. These may not have been gastropods at all, and many only bear a coincidental resemblance to the true Bellerophontacea.

Helically coiled shells are first seen in the U. Cambrian. They are archaeogastropod genera such as *Pleurotomaria* and *Maclurites*, examples of the former being still alive today. In the Ordovician, the caenogastropods appeared. They were restricted to the oceans, possibly until the Carboniferous, when they may have radiated into fresh water.

But the most successful gastropods of both land and fresh water, the pulmonates, did not appear until some time in the Mesozoic. After this, during the Tertiary, gastropods underwent their greatest expansion — accompanied by the evolution of the pteropods (Subclass Opisthobranchiata) some time in the Eocene.

The success of gastropods in the Tertiary has led to their use in the stratigraphy of Tertiary rocks, especially in areas (such as Britain) where, for much of the time, sedimentation was freshwater or shallow marine. So, although the base of the Eocene Barton Beds is marked by the appearance of the foraminifer *Nummulites* (Ch. 10), the subdivision of the rest of the unit relies upon the molluscan fauna, which consists chiefly of gastropods.

But although they are of rather limited stratigraphic value, gastropods are unrivalled in the identification of environmental conditions from the more recent past. They have been of immense value to archaeologists (section 1.11) concerned with the environment of early man. Many of the snails then living are still alive today, and are known to have very strict preferences in terms of climate, rainfall, type of vegetation and so on. Not only have they enabled us to document changes in climate since the last glaciation, but we have even been able to chart the effects of man's activity upon the vegetation, as he moved into possession of the greater part of post-glacial Europe.

74 *Palaeontology — An Introduction*

II. CEPHALOPODS

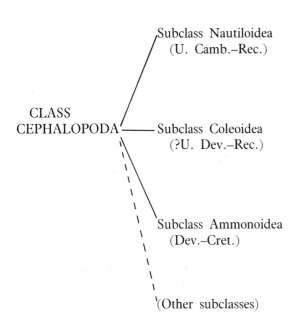

CLASS CEPHALOPODA

Subclass Nautiloidea (U. Camb.–Rec.) — External shell, though some may have been internal. Mostly straight cones, though *Nautilus*, a living representative, is coiled. Most Palaeozoic cephalopods are nautiloids.

Subclass Coleoidea (?U. Dev.–Rec.) — Internal shell. Includes all living cephalopods except *Nautilus*. Squids, cuttlefish, octopuses. Belemnites notable fossil forms.

Subclass Ammonoidea (Dev.–Cret.) — External, mostly coiled shells. Prominent in Late Palaeozoic and Mesozoic. Goniatites and ammonites most notable fossils.

(Other subclasses)

6.13 Soft parts and life cycle

Cephalopod anatomy is strikingly different from the basic molluscan plan. In general, there has been a 90° shift of the standard orientation, from a ventral head-foot with a dorsal shell, to a shell borne posteriorly and the head-foot pointing forwards (Figs 6.1, 6.7). The foot has also been modified into many **tentacles**, and

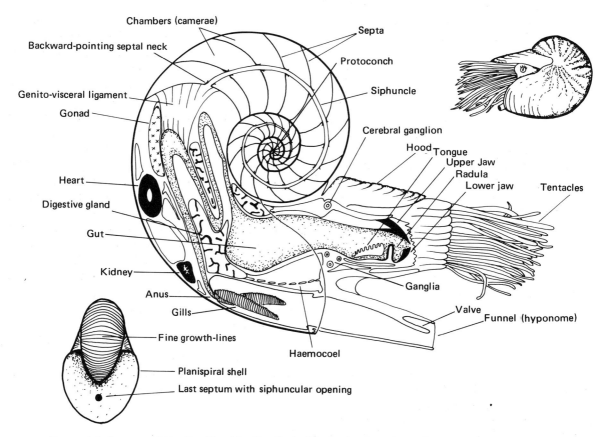

FIGURE 6.7. Anatomy of *Nautilus*. *Top right*: *Nautilus* in life showing the pattern of coloration on topside of shell. *Bottom left*: Apertural view of empty shell, showing last septum and its siphuncular hole. (After Moore, Stenzel and others.)

the mantle cavity — having migrated all the way from the back — has come to point anteriorly. Its opening is surrounded by a muscular organ called the **funnel**, which can direct jets of water in any direction — so controlling the characteristic 'jet propulsion' mode of locomotion in this group.

The head projects into the ring of tentacles, bearing a mouth and (usually) a pair of highly efficient eyes. The mouth often carries a pair of jaws shaped like a parrot's beak, and the shell, which may be a straight, curved or coiled cone, is divided into **chambers** by walls called **septa**. Its function, when worn exteriorly, is partly protective, since the soft parts may withdraw into it in emergencies. However, in all cases it functions like a submarine's ballast tank in controlling the buoyancy of the animal.

The sexes are separate in cephalopods, and complex courtship rituals may precede copulation. This involves the placing of a ball of sperm into the female's mantle cavity by means of one or many specialized, erectable tentacles carried by the male. Such differences in soft parts between males and females may be mirrored in their hard parts, so that fossil males of one species may look very different from the females (section 6.18).

Since cephalopods are free-swimming, they have no need of larval stages and so develop more or less directly from the egg. They can escape rapidly from danger by the strenuous contraction of the mantle cavity, which forces an extra-high-speed jet of water from the funnel (or **hyponome**, as it is also called). In squids, this is accompanied by the release of ink from a sac just behind the mantle. Beneath the cover of his dense cloud the squid can then escape. Fossil ink sacs have been found not only among fossil coleoids but also in some Jurassic and Cretaceous ammonoids.

A biologist, dealing with living cephalopods, can recognize two major groups. One of these has four gills in its mantle cavity (*Nautilus*, for example), but the other only two (e.g. squids). Despite its convenience for living forms, this scheme is obviously inapplicable to fossil groups — especially those with no living representatives. Here, therefore, we have the opposite state of affairs to that seen in gastropods; here, palaeontologists are unable to apply a biological classification because its anatomical distinctions are not reflected in any skeletal feature.

6.14 Shell form

The shells of cephalopods are borne in two ways. Either they are carried internally, completely enveloped in soft tissue, or they are external, like the shells of most other molluscs. As a rule, external shells are found in the Nautiloidea and Ammonoidea, while internal ones are characteristic of the Coleoidea. But it should be noted that the basic elements of cephalopod shell design are present in both internal and external shells.

All cephalopod shells are chambered. This is because the animal, as it expands and grows, lengthens and widens its shell and moves towards the ever-advancing aperture. This leaves a space behind the body, which is periodically walled off by a **septum**. The septa join the shell along a **suture line**. Because in many forms the septum is not a simple, flat structure but a highly folded one, this suture may be extremely sinuous.

Sutures, such as those displayed in Fig. 6.9, are very important indeed to cephalopod classification. Remember that they are the lines of attachment of the septum to the shell *interior*, and are not expressed on the exterior surface. Neither do they bear any relation to growth lines or any external ornament which may be present. Fossil forms commonly display sutures because they are preserved as internal moulds, with the shell dissolved away.

So, unlike the gastropods, cephalopods do not inhabit all of their shell volume. Their bodies are confined instead to the outermost chamber or **body chamber** (Fig. 6.7). This is structurally rather weak, and so is often destroyed before preservation in the fossil state. Many fossil cephalopods are therefore most frequently represented by the septate portion of the shell only. This portion is called the **phragmocone** (Fig. 6.8).

The living animal in the body chamber maintains contact with the chambers of the phragmocone by means of a cord (the **siphuncle**) which travels from its rear end right back to the first septum, near the origin of the shell. Every septum is therefore perforated by a hole (the **foramen**) through which the siphuncle passes. This hole may be in the middle of the septum or at its edge near the shell margin, or anywhere in between. It may also be projected into a small tubular sheath which in life surrounds the siphuncle. This sheath may project either forwards (towards the aperture) or backwards, and is another taxonomically important feature.

The characteristics described so far may be seen in nearly all cephalopod shells. Look for them in the diagrams showing representatives of each of the three subclasses discussed in this chapter (Figs. 6.8, 6.10). The diagnostic characters of ammonoid, nautiloid and coleoid shells are shown, together with their technical nomenclature.

The next part of this discussion of shell form explains some of the more noteworthy aspects of form which are to be encountered in the externally shelled groups (Ammonoidea, Nautiloidea).

6.14.1 Cameral deposits

In straight-shelled nautiloids (Fig. 6.10), heavy deposits of calcium carbonate were formed, beginning

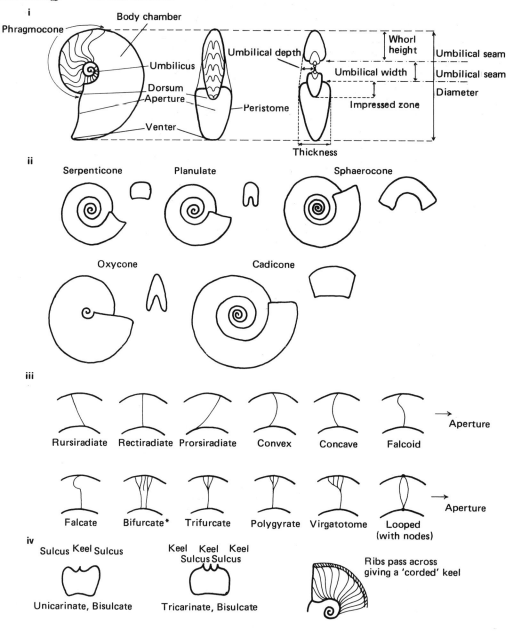

FIGURE 6.8. Shell morphology and ornament. (i) Terms of reference for ammonoid shells. (ii) Styles of coiling and whorl-section. (iii) Styles of ribbing seen in ammonites (aperture to right in all cases). (iv) Types of marginal ornament. Keels, sulci and corded keel.

at the apical end and advancing, as the shell grew, into progressively younger chambers (or **camerae**, as they are also known). Apical chambers were frequently completely filled in this way. Deposition tended to be heaviest on the ventral sides of chambers (Fig. 6.10 ii). The function of these deposits is discussed in section 6.17 below.

6.14.2 Coiling

Ammonoids and some nautiloids commonly show planispiral coiling. Such shells never develop cameral deposits.

It is useful to describe such shells in terms of the 'tightness' of the coil, which amounts to the degree to which older whorls are embedded in the younger ones. If the coils are only touching or only slightly embedded, the spiral is an open one and all the older whorls remain visible. Such shells are termed **evolute**. In **involute** shells, the last whorl may overlap almost all the phragmocone, leaving a very small depression at the centre of the shell.

Varying degrees of involution combined with differing whorl sections can produce many shell morphologies in coiled cephalopods. Some of these are shown and named in Fig. 6.8.

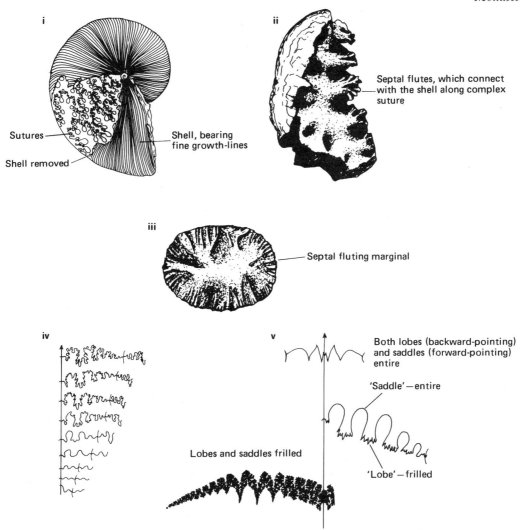

FIGURE 6.9. (i) *Phylloceras heterophyllum*. External shell with fine growth-lines removed in part to reveal sutures. U.Lias, Whitby (Yorks.) Redrawn from 'British Mesozoic Fossils' (British Museum). (ii) The septal surface of *Placenticeras* (Cretaceous, Alabama, U.S.) showing how the septal fluting dies out inwardly from the suture (drawn from a photograph in Kennedy and Cobban, 1976). (iii) Septal surface of *Baculites*, a straight-shelled Cretaceous form (redrawn from Fenton and Fenton, 1958) (iv) Sutures of *Creniceras reuggeri* for one side of the shell only. This is the standard means of representing sutures. Those from the early shell are at the bottom — the arrow points to the aperture, and its line represents the mid-venter. The curved lines represent the edge of an overlapping whorl which had to be removed to reveal all of the suture. Note how the complexity of folding increases toward the aperture. (From Palframan, 1966). (v) Some representative sutures. Top, a typical goniatite *Goniatites crenistra* (Viséan). Middle, a typical ceratite suture (O. Ceratitida). Only the backward-pointing loops are frilled. Bottom, a complex ammonite suture (O. Ammonitida) with both lobes and saddles frilled.

6.14.3 Shell aperture

Although the shape of the shell aperture is very much dependent upon the tightness of coiling, it may also be affected by notches and/or projections. Quite commonly there may be a notch mid-ventrally. This allows the funnel greater freedom of movement and so improves mobility. It is known as the **hyponomic sinus** (Fig. 6.10). In some forms (see section 6.18) there may also be two lateral projections on either side of the aperture, called **lappets**. They probably served as copulation aids in male ammonites (Fig. 6.12).

6.14.4 Opercula

Plates which closed the aperture when the animal retracted into its shell are known, but are rare in cephalopods. Like gastropod opercula (section 6.6) they were attached to the soft parts and so became lost upon decay, in most cases. Perhaps the best known are ammonite **aptychi** (Fig. 6.12): two calcitic plates which together fitted the aperture perfectly, and are occasionally found in position. It is interesting to note that they were calcitic, while the ammonite shell itself was made of aragonite. Why this should have been so

78 *Palaeontology — An Introduction*

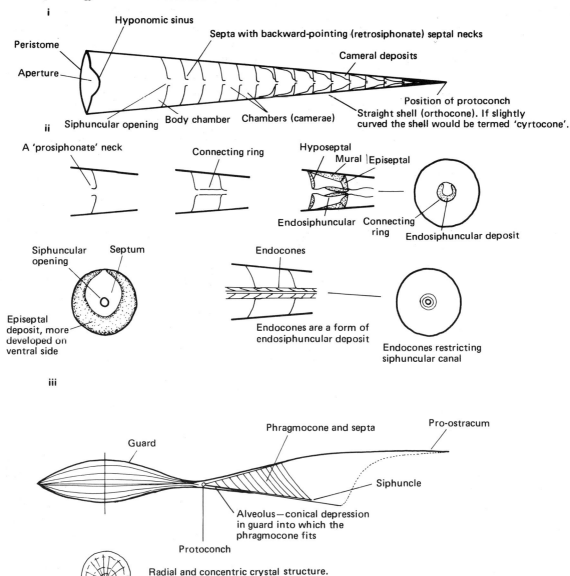

FIGURE 6.10. (i) Orthocone morphology. (ii) Detailed morphology of septa and cameral deposits. Many orthocones have septa linked by connecting rings which surround the siphuncle. These 'rings' may be parallel-sided cylinders as shown above, but may also bow outwards to form bulbous or inflated structures. Within the connecting rings deposits may be laid down. One special type of endosiphuncular deposit are the endocones, which stack like conical paper cups inside the siphuncular canal and considerably restrict its diameter. Terms for different types of cameral deposit are also shown. In older regions of the phragmocone cameral deposits may fill chambers completely. Cameral deposits tend to be laid down most thickly on the ventral sides of chambers. (iii) Belemnite morphology. The soft parts of belemnites extended over the entire shell enabling the secretion of the guard on the posterior end. You can see that a belemnite is basically an orthocone with additions front and back.

is not known for sure, but it means that aptychi are often well preserved.

6.14.5 Ornament

External ornamentation of the shell was very varied, especially in ammonoids. It is very useful in description, though its precise function is not always easy to deduce. The variety of fine striations, strong ribs, wide flanges, knob-like tubercles and extended spines was very great indeed. Most were produced by folding the shell, rather than by local thickening. This means that they are expressed on internal moulds, which is very fortunate.

Notice also that the periphery of the shell may also have been adorned with ridges (**keels**) and grooves (**sulci**). The ridges themselves may also have borne fine ornament (Fig. 6.8 iv).

6.15 Belemnites

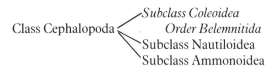

Although the Subclass Coleoidea contains almost all living cephalopods, only the belemnites (Order Belemnitida) are important in the fossil record. Like other coleoids, their shells were covered with living tissue. We shall consider them in their three component parts (Fig. 6.10 iii).

Most commonly found is the posterior portion, a heavy calcitic, bullet-shaped object called the **guard**. It ends posteriorly in a point, while at its other end it has a small conical depression called the **alveolus**. Into this fits the end of a delicate, chambered **phragmocone**. This is very similar to the phragmocone of a straight-shelled nautiloid, but it has a ventral siphuncle. It may extend out beyond the alveolus, and was originally aragonitic. This is one reason why phragmocones are less frequently found than guards, although mechanical robustness is, of course, another factor (see also section 6.19).

The third and last portion is the most anterior of all, and takes the form of a hood. Its delicacy is such that it is extremely rarely preserved. It is called the **proostracum**, and it may be regarded as an incomplete body chamber.

We can tell by the way the guard appears to have been 'stuck on' to the back of the phragmocone that living tissue must have extended right over the shell in belemnites. Many different shapes of guard are seen, but nearly all of them have the same, streamlined, torpedo shape. This tells us that, like modern cuttlefishes, the direction of motion (at least during escape manoeuvres) was backwards, guard-first. The function of the guard is discussed in section 6.17 below.

Belemnites can be very easy to find, especially when the rocks which contain them are soft, and the fossils can be washed out by the weather or by wave action.

6.16 Ammonites

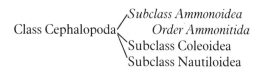

The three subclasses with which we are concerned are each large groups, comprising many orders. And just as the belemnites compose one order of the Coleoidea, the ammonites make up the Order Ammonitida of the Ammonoidea. They are so important in cephalopod palaeontology that they require separate treatment.

Ammonites dominated the Mesozoic cephalopod fauna, evolving so rapidly and becoming so plentiful, diverse and widespread that they are today the ideal stratigraphic tool. Another ammonoid order, the Goniatitida, underwent a similar but less spectacular radiation in the Permo-Carboniferous, in which systems they too are stratigraphically important. They are discussed further in section 6.19.

What distinguishes ammonites from other members of this subclass — the goniatites, for instance? Like all ammonoids, the ammonites tended to have coiled, planispiral shells and ventral siphuncles. Careful medial sectioning would also reveal a bulbous **protoconch** at the origin of the spiral.

But ammonites in particular had extremely elaborate suture lines. Look at the examples of the typical sutures seen in various cephalopod groups (Fig. 6.9 v) and you will see that the ammonite suture is the most complex of all. Another tendency among ammonites is for the **septal necks** to point forwards, while on the outside of the shell an extremely wide range of shell ornament may be encountered.

Nearly all post-Triassic ammonoids are members of the O. Ammonitida, so you can be fairly confident that any ammonoid you may find in Jurassic and Cretaceous rocks is actually an ammonite!

6.17 Mode of life

The environmental stresses with which aquatic animals must deal come in two categories. **Hydrodynamic** effects operate when fluid is in motion (or when the animal moves through the fluid). **Hydrostatic** effects, on the other hand, are due to the properties of fluids at rest — increase of pressure with depth, for example.

We are all familiar with these different properties, for when we swim across a pool the resistance of the water against our bodies is a phenomenon of hydrodynamics. If we dive to the bottom, the pressure we feel as the water compresses the air in our lungs is a phenomenon of hydrostatics.

The hydrodynamic features which adorn the bodies and shells of marine animals may often be quite easy to identify; we have already mentioned the shape of the belemnite guard, designed to allow for ease of movement through the water during rapid retreats. Similarly, a submarine is shaped for hydrodynamic efficiency, but it also possesses great strength under compression, its many bulkheads supporting the external skin. These bulkheads are hydrostatic features.

A submarine also has ballast tanks which, by flooding or emptying, can be used to change the ship's buoyancy. When empty, the density of the ship as a whole is less than that of water, and it floats. When full, the average density is greater than that of water, and it sinks. Somewhere in between, with tanks partially flooded, is a condition of neutral buoyancy, when the submarine will remain suspended in the water column.

Animals which live at a wide range of depths need, like the submarine, to control their buoyancy. They have to be able to rise and sink at will, and to float in the correct attitude, just as the submarine has to stay more or less horizontal in the water to be able to function properly.

Let us look now at *Nautilus*, and see how it employs its shell to do these things. We can extrapolate our observations from *Nautilus* to fossil coiled cephalopods. Maybe an appreciation of hydrostatics will help us to explain many of the puzzling features of shell form in cephalopods.

In living *Nautilus*, the chambers of the shell contain gas and, in the younger chambers near the aperture, a certain amount of liquid (**cameral fluid**). The pressure in these chambers tends to be at just less than atmospheric pressure, which means that at depth the shell must withstand considerable compressive stress.

The buoyancy of *Nautilus* is adjusted by the addition or removal of fluid in the chambers via the siphuncle. As in the submarine, flooding of the chambers allows the animal to sink, while withdrawal has the opposite effect. The system works efficiently enough to allow the animal to rise in the water column, come inshore to feed during the night, and to return to depths of up to about 600 m by day.

Of course, the disadvantage of the system is the danger that the shell might implode at depth, for it is under immense pressure. But experiments have shown that the simple lines and perfect design of *Nautilus* convey sufficient strength to allow for a pressure-tolerance limit well above the range normally demanded by the habits of the animal.

Ammonite shells seem to have possessed all the necessary structures for them to have functioned as hydrostatic devices, just like the shell of *Nautilus*. But what of their complex sutures, their much more elaborate ornament, or the great variation in the shape of the shell as a whole?

Firstly, it is clear that when a chambered shell is under pressure the weakest point of the surface is that which is furthest from any form of support. So if there is a real danger of implosion there will therefore be considerable advantage in producing a shell design which minimizes the amount of unsupported external shell.

In the cephalopod design, one way of doing this is to fold the septum near its contact with the shell. This lengthens the line of contact between the septum and the shell, so that the compressive force borne per unit length of suture is reduced. Moreover, the amount of unsupported shell surface is also cut dramatically. So we can view complex sutures as being of adaptive advantage in shell strengthening. (See Chapter 12 for further analysis of this problem.)

In ornamented shells, ribbing may also have had a strengthening role, but strength under compression is more likely to be maintained, as in *Nautilus*, by clean, simple lines. Some ribs may have had hydrodynamic properties, since it has been noticed that, under certain conditions, they actually *reduce* water resistance. Some may have acted like ships' stabilizers, to prevent pitching and rolling. Other features may only have functioned when they were sited at the aperture, having no special purpose thereafter.

Many marine animals are camouflaged. *Nautilus* is striped on its upper surface (Fig. 6.7) so that its outline, when seen from above, is broken up. When viewed from below against the brightness of the water surface, its white underside reflects light and makes its silhouette less prominent. Many fish also employ this system of camouflage — look at a mackerel the next time you visit a fishmonger.

Many of the knob-like tubercles or extended spines of ammonites could have been present to diffuse the shell outline in a similar way, possibly enhanced by pigmentation. Original colour is rarely preserved in fossils, but there are known cases of Palaeozoic orthocone nautiloids possessing stripes and zig-zag markings on their dorsal surfaces.

Such markings tell us that the life position of orthocones was horizontal. But how did they maintain their orientation? Think of the distribution of density in an orthocone nautiloid. All its chambers were empty, while at the anterior end was a relatively dense mass of living tissue. One would think the tail-end would have tended to rise buoyantly, so tipping the creature vertically.

It was, in fact, to prevent this very thing that cameral deposits were laid down. They were precipitated most intensely in the posterior chambers (where their counterbalancing effect would have been greatest) and also predominantly on the ventral sides of those chambers. Their effect counteracted the tendency of the shell to tip up, while they also prevented rolling along the long axis by being concentrated ventrally.

Now we can see that the belemnite guard was merely another solution to the problem of staying horizontal. Belemnites, however, could put their counterweight on the outside of the phragmocone because of their external tissue layer.

When gravity acts upon a body, the forces of attraction exerted by the earth act about a theoretical point known as the centre of gravity. This is not necessarily in the geometrical middle. Consider the geological hammer; because most of its mass is in the head, its centre of gravity lies towards that end.

By analogy, we can think of the forces of buoyancy as acting upon a theoretical point in any floating object, called the centre of buoyancy. In the geological hammer, this would fall somewhere in the wooden shaft.

Applying these ideas to the orthocone (Fig. 6.11 ii),

Molluscs 81

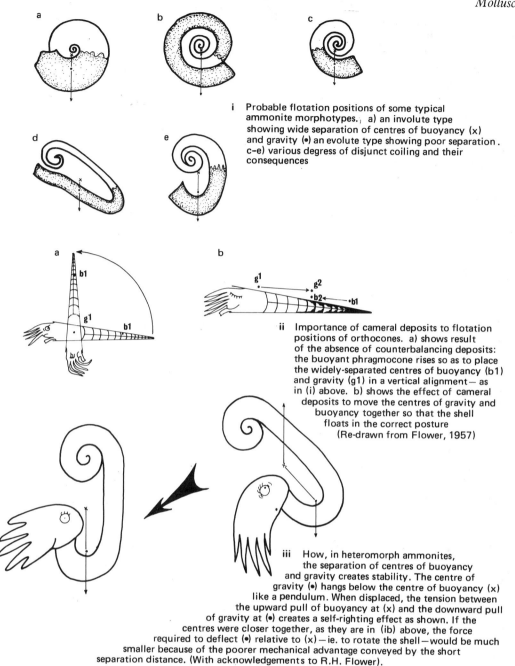

FIGURE 6.11.

we can represent the effect of putting cameral deposits in the apical chambers in terms of shifting the centres of buoyancy and gravity towards each other so that they are actually in the same place. Without the deposits, the centre of gravity lies at the anterior end of the shell, with the soft parts. The centre of buoyancy lies in the area of the buoyant phragmocone. The opposing forces thus turn the shell into a vertical position where there is a stalemate, a straight-line tug-of-war between buoyancy and gravity.

Coiling the shell into a plane spiral (the usual form of ammonoids, but seen also in *Nautilus*) allows the centres of gravity and buoyancy to remain continuously in this vertical tug-of-war arrangement (Fig. 6.11 i).

This eliminates the need for cameral deposits and is a very successful solution to the problem of maintaining a correct attitude in the water.

In coiled shells, we find that different modes of coiling produce different degrees of stability. Evolute shells, like that of *Dactylioceras*, have their centres of buoyancy and gravity rather close together. Involute shells, in which the body chamber was much shorter, manage to attain considerable separation.

In the orthocone the imperative was to have the two centres *coincident*. But with coiled shells the opposite is true if the animal needs to move efficiently. Remember that cephalopods propel themselves by firing jets of water from the funnel in the manner of a jet or

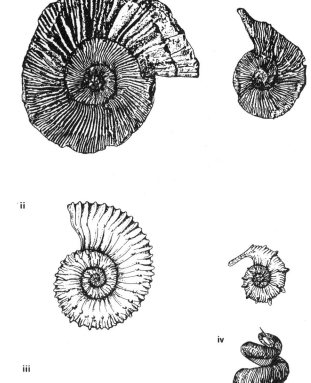

FIGURE 6.12. (i) Sexual dimorphs. Macroconch (*left*) and microconch of *Pectinatites reisiformis* (Jurassic). Note the rostrum of the microconch. (Drawn from photographs in Cope, 1967.) (ii) Sexual dimorphs, originally described as different species; Macroconch (*left*) *Cosmoceras spinosum*, microconch (with lappets) '*Cosmoceras annulatum*'. In this and other such cases the older name takes precedence and is used to refer to both forms. (Redrawn from Makowski 1962.) (iii) Ammonite aptychi showing fine growth lines on inner surfaces. *Laevaptychus latus* (Jurassic) Bavaria. (Drawn from a photograph in Kennedy and Cobban, 1976). (iv) A heteromorph ammonite *Didymoceras nebrascense* (Cretaceous) Colorado. (Redrawn from Kennedy and Cobban, 1976.)

rocket engine. If the form whose centres of buoyancy and gravity lie close together were to give a sharp burst of thrust, much of the force would result in the animal spinning round (like the Catherine wheel it resembles).

Separation of the centres prevents this. The distance over which the tension between them acts, creates stability by virtue of the mechanical advantage offered by the leverage which that distance creates. Like a pendulum, the shell requires considerable force to rotate it about its 'pivot', and it will always swing back afterwards into the correct position.

Attainment of this kind of stability appears to have been the motive behind the dramatic examples of uncoiling seen in the so-called heteromorphic ammonites (Figs. 6.11 iii, 6.12 iv). We may deduce from this that these, and the involute types, were efficient swimmers, while evolute shells may have belonged to more slowly moving, perhaps benthic, animals. In such a mode of life their deficiency as swimmers might not have been so exposed.

6.18 Sexual dimorphism

We cannot help but be very much aware of the physical differences between males and females of our own species. But, without the soft parts available, would some palaeontologist of the future, studying extinct humanity by its fossil bones, be equally aware of them?

The answer is yes, because the proportions of male skeletons and female ones are different. To prove this to yourselves, you can make a graph of height versus hip girth from a sample of sexually mature males and females of normal body weight. You should find that males (who tend to be taller and have narrow hips) plot in a distinct cluster from females (who are generally less tall and have broader hips).

The expression of gender in body form is called **sexual dimorphism**. The degree to which it occurs can vary enormously from species to species. In one species of Anglerfish, *Ceratias holbolli*, the male is about 4 cm long and spends his life attached permanently to the female, who is 75 cm in length. The male is little more than a sperm-producing organ of the female's body, having lost all independent existence — even down to fusing his vascular system with that of his mate.

Disparities on this scale make our own seem rather trivial, but sexual dimorphism of some kind is widespread, and its function differs from case to case. The human female's broad hips, for example, serve to support her abdomen during pregnancy. Human males tend to be larger than females, and the same is true of many mammals. But in the animal kingdom as a whole, the reverse tends to be the case. This is, simply, because the production of offspring requires a larger biological machine than the production of sperm.

So, if living animals are a reliable guide, we might expect fossil species to consist of two forms, not one; there should be a male form, and a female form. This is a disturbing thought, because fossil species have been erected on criteria of form alone, and so it is likely that all sexual dimorphs will, in the past, have been wrongly split into two separate morphospecies.

How, then, are we to go about recognizing sexual pairs and reuniting those which the taxonomist may have put asunder? Here are some guidelines.

1. Geological range. Obviously, one form could not have survived without the other, so the two forms should appear in the fossil record, and disappear from it, simultaneously.
2. The two forms should be found together in the same (geological) bed.
3. No 'intermediate' forms should exist. The two morphologies should be quite distinct from each other, without gradation.
4. Sexual differences only develop at sexual maturity. In ammonites, therefore, the early whorls should be identical, and differences confined to the later growth stages.
5. The numerical ratio of one form to another should be approximately 1 : 1 (though this is actually rarely the case — see below).

Work on ammonite dimorphism has revealed that the males and females were strikingly different in size. The smaller form (termed **microconch**) reached its full size early in life, tended to maintain its juvenile ornament unmodified, but commonly developed lappets (section 6.14, Fig. 6.12).

The larger form, or **macroconch**, usually grew at least one extra whorl before attaining its full size. It also tended to reduce its ornamentation. It never developed lappets.

So, following a general rule-of-thumb, it is generally assumed that the macroconch was the shell of the female form and the microconch that of the male. As has been said, the microconch's lappets (or sometimes, other kinds of apertural prolongation) could have had some copulatory function.

Palaeontologists have found that the ideal, 1 : 1 ratio of macro- to microconchs is rarely attained. Happily, we can partially account for this by examining living cephalopods: for in some populations the sex ratio is naturally unequal (usually with a preponderance of females). Moreover, many species may spend much of their lives in monosexual shoals.

Lastly, the fact that ammonites were often dimorphic raises the interesting possibility that much of their ornament may have had some indirect or 'secondary' sexual purpose. Like all cephalopods, ammonites probably had good eyesight. This may lend support to the idea that their exterior decoration could have functioned in sexual display and attraction.

6.19 Geological history and stratigraphic value

The Subclass Nautiloidea includes the oldest known cephalopod *Plectronoceras*, which appeared in the late Cambrian and has a simply curved (cyrtocone) form. As with many fossil groups, the nautiloids enjoyed a marked radiation in the early Ordovician, and nearly all component orders of the subclass had evolved by the middle of that period — all except the Order Nautilida, which contains our modern *Nautilus*. This order originated at about the Silurian/Devonian boundary.

Despite its late start, it was the Nautilida which gave rise to all post-Palaeozoic nautiloids: nearly all of which were planispirally coiled. At present, however, the subclass is in decline, represented only by a few species of *Nautilus* in modern oceans. With only these standing between it and extinction, it would appear that a minor recovery which it staged in the early Tertiary has come to nothing.

The Ammonoidea evolved in the Devonian, deriving from one of the many minor subclasses which we have not discussed in this book. This minor subclass comprised cephalopods which had what we now refer to as typical 'ammonoid' characteristics (bulbous protoconch, ventral siphuncle) while at the same time possessing a simple **orthocone** shape.

This began to change in the early Devonian, when some representatives began to curl (become **cyrtocones**) and eventually to coil. An evolutionary sequence led, by a series of ever more strongly coiled intermediates, to the gradual production of the planispiral form. The Order Ammonoidea had been born — and after the mass extinctions at the end of the Permian, its surviving orders underwent great expansion (Fig. 6.13).

Before this great Mesozoic radiation, however, Ammonoids did become quite important members of the late Palaeozoic fauna. One order, the Goniatitida, evolved rapidly enough, and attained sufficiently wide distribution in the Devonian and Carboniferous, for them to be used today as zone fossils.

For example, at the end of Carboniferous Limestone times in Britain, shallow marine carbonate deposition gave way to fluvio-deltaic sedimentation. In South Wales, a rapidly fluctuating shoreline was created by many minor transgressions, which brought marine faunas in for brief periods over previously estuarine and deltaic environments. These 'marine bands' contain goniatites, and are named after characteristic forms. Their lateral persistence allows the successful correlation of the very variable deposits which lie in between, and the topmost band, the *Gastrioceras subcrenatum* Marine Band, marks the internationally agreed base of the Coal Measures over the whole of Europe.

In the Triassic, after the great extinctions of the preceding period, nearly all surviving ammonoids belonged to the Order Ceratitida, most of which displayed a distinctive suture type (Fig. 6.9 v). From these evolved another order, the Phylloceratida, which was to give rise to all the post-Triassic ammonoids when the ceratitids finally became extinct (Fig. 6.13).

The Phylloceratida, and their daughter order the Lytoceratida, together acted as root stocks for all the superfamilies which came to make up the Order Ammonitida. Each stock remained more or less unchanged throughout, periodically 'budding off' a sub-group. This pattern of evolution, with a main stock continuously branching sideways but remaining intact itself, is called **iterative evolution**. Also, you will notice that the Ammonitida probably derived from *two* ancestral groups, not one. It is therefore a **polyphyletic** group.

Ammonites are the prime stratigraphic tool in the Mesozoic. With their wide distribution, rapid evolution and almost perfect facies independence, the Jurassic and (to a lesser degree) the Cretaceous can be zoned with a precision which, in some cases, narrows down to less than half a million years. Their only shortcomings are their absence from non-marine rocks and — for some unknown reason — from reefs. They are also of little use to the oil industry, for reasons explained in Chapter 10.

Diversification in ammonites continued unabated until they were finally extinguished at the end of the Cretaceous, in another round of mass extinction. In this instance, not only the order was wiped out, but with it the entire subclass.

The use of belemnites (Subclass Coleoidea) as zone fossils is hampered by a paucity of identifiable features, rarity of complete specimens and (from the human side) a dearth of specialist workers interested in them. Nevertheless, some belemnites, such as *Arcoteuthis* and *Hibolites*, characterize rocks of the L. Cretaceous, while one may find *Gonioteuthis*, *Belemnitella* and *Belemnella* in the U. Cretaceous chalk — known sometimes as 'the Belemnite Chalk'.

One problem with belemnite guards, however, is their astonishing durability. Because of this they may easily survive re-erosion from their native horizon and be redeposited in another — leading to the danger of mistaken correlation. Such derived belemnites are known from the basal conglomerates of the Pliocene in East Anglia.

But durability can have its advantages, for even the highly metamorphosed and strongly deformed *Schistes Lustrés* of the Pennine Alps have been successfully dated by their (still recognizable) belemnite fauna!

III. BIVALVES

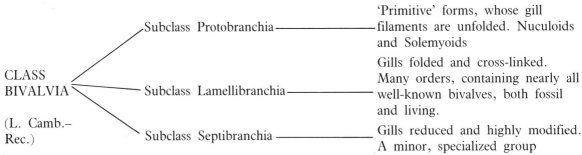

[Note: Although 'Bivalvia' is the original name of the Class, given by Linnaeus, bivalves have been referred to as 'pelecypods' and 'lamellibranchs'. The former term is not employed here. The latter is taken to refer only to the major subclass, detailed above. Note also that the system of classification is a 'zoological' one, based on soft-part anatomy (gill morphology). Other 'palaeontological' schemes have been put forward, but they are not considered here.]

6.20 Soft parts and life cycle

In bivalves the soft parts (Fig. 6.19) are partially or completely enclosed between two hinged **valves**. Most of the viscera are to be found close to the back of the enclosed space, adjacent to the **hinge**. A **mantle** lines both valves almost as far as their outermost edges, and is responsible for secreting the shell.

The body does not occupy all the space inside the valves even when closed. The residual space is taken up by the **mantle cavity**, into which hang the respiratory **gills** and the muscular **foot**. The gills create currents using the **cilia** with which they are covered, drawing in water which, typically, they also strain for food particles. These are then conveyed in a stream of mucus to the mouth.

The foot is shaped rather like the head of an axe, and can be protruded from the shell to aid in burrowing (Fig. 6.19 iii). Most bivalves live at some depth below the sediment surface, but some lie on the sea bed, or attached permanently to rocks or seaweed. In these latter cases the foot loses its usefulness and tends to become atrophied.

In burrowing forms, the need for a constant supply of water for food and respiration is satisfied by the use of snorkel-like tubes (**siphons**), one inhalent and one exhalent. In most bivalves these are formed by the mantle edge, which fuses along part of its length. Those unfused portions are drawn out into fleshy hollow cylinders which extend to the surface. Fusion of the mantle also excludes sediment from the mantle cavity, but in most burrowers there is also a third gap, left open to allow for the protrusion of the foot.

The shell is closed by the action of **adductor**

Molluscs 85

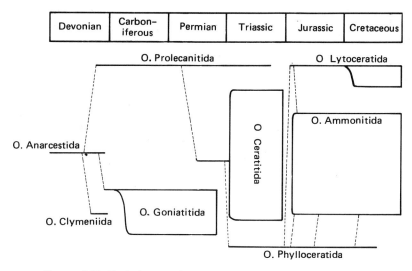

FIGURE 6.13. Evolutionary relationships between the orders of the Class Ammonoidea (modified after Moore in 'Treatise' (pt. L.))

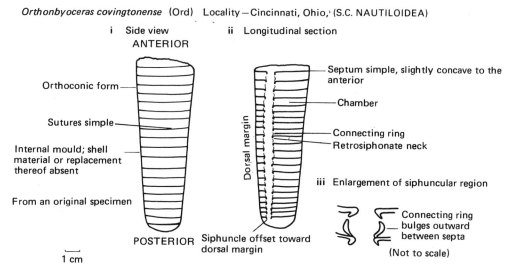

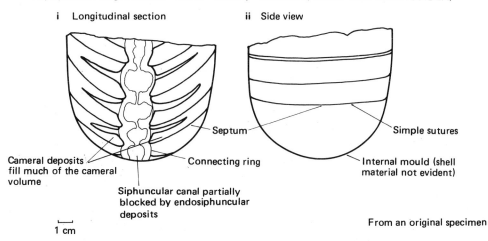

FIGURE 6.14.

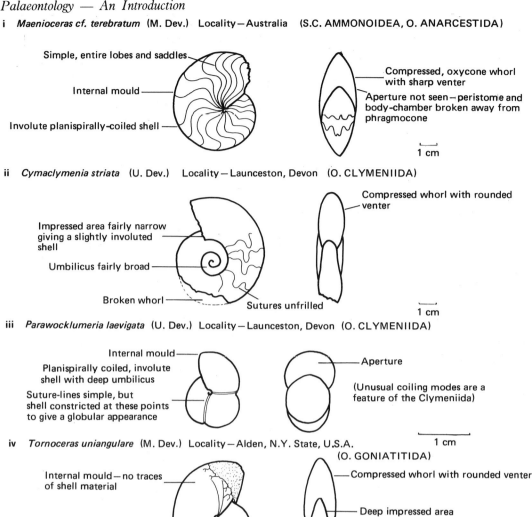

FIGURE 6.15. Anarcestid, clymeniid and goniatitid ammonoids. (All drawn from original specimens.)

muscles which connect the valves. There are usually two of these and the impressions (**scars**) which they leave on the shell interior allow the musculature of fossil forms to be reconstructed. In addition, one may, in some shells, see two other scars, produced by the muscles used to withdraw the foot. They tend to lie on the dorsal side of the adductors and are called **pedal retractor muscles**.

We have seen how brachiopods (Chapter 5) employ muscles to close and to open their shells. Bivalves have a different system. Their shells are permanently spring-loaded, so that they have a constant tendency to open. If you find an articulated bivalve shell on the beach, from which the soft parts have been removed by decay or scavengers, you will notice that they always gape open. It may require some force to close them.

This constant opening force is applied by a structure called the **ligament**, which is described in the next section. Although more resistant to decay than the body, it too eventually fails and the two valves commonly fall apart. This is because, unlike most brachiopods, the dentition of bivalves does not lock the valves firmly together. It is therefore relatively rare to find articulated specimens, unless they have been preserved in life-position.

In moving from the generalized molluscan plan through the gastropods to the cephalopods, we have

i *Stenopronorites arkansasensis* (U. Carb.) Locality—Clarita, Oklahoma, U.S.A.
(S.C. AMMONOIDEA, O. PROLECANITIDA)

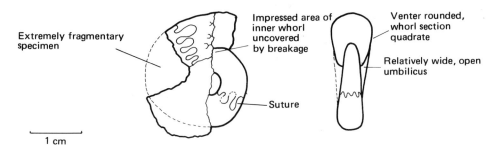

ii *Ceratites nodosus* (M. Trias) Locality— unknown (O. CERATITIDA)

iii *Arcestes sp.* (U. Trias) Locality — unknown (O. CERATITIDA)

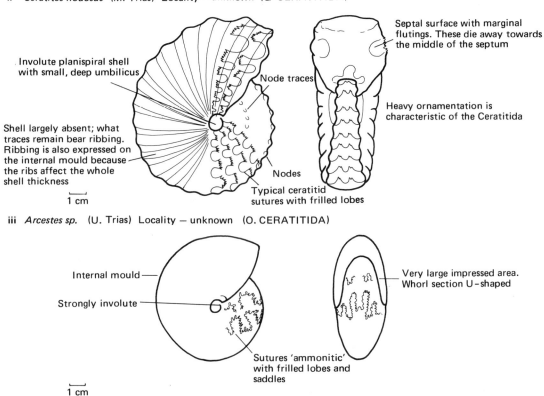

FIGURE 6.16. Prolecanitid and ceratitid ammonoids. (All drawn from original specimens.)

seen the development of predatory life styles in the latter to be matched by (a) increasingly efficient eyes, (b) a better-developed nervous system for improved coordination, and (c) the concentration of sensory organs and mouth into a recognizable 'head' region.

In the bivalves, on the other hand, the filter-feeding existence has had the opposite effect. A filter feeder need not look for its prey, and little coordination is required for sluggish life in a deep burrow. The result has been the loss of the 'head' and the total reduction of eyes — though some, like the scallop *Pecten*, have secondarily redeveloped 'eyes' all around the mantle edge. (Here, incidentally, is another example of the irreversible nature of evolution, touched on in section 6.10 above. Just as the pulmonate gastropods 're-invented' the gill when they became adapted for aquatic life, the *Pecten* developed a visual system independently, owing nothing to the primitive molluscan eye (Fig. 6.1), which in bivalves has been completely evolved out.)

The mouth receives mucus-bound food particles from the gills via the **palps**. These sort the food, conveying some to the mouth but rejecting others. The gut ends at an **anal pore** within the mantle cavity, where faeces are released and carried out in the exhalent stream.

Bivalves, with their fixed or burrowing life styles, are clearly not suited to sexual reproduction involving copulation. For them, as for brachiopods and most sessile (static, fixed) organisms, the task of interchanging genetic material is left to motile sex cells released into the water. Dispersal is achieved by the larval stage which forms upon fertilization. This may live for several months, during which time it seeks a suitable

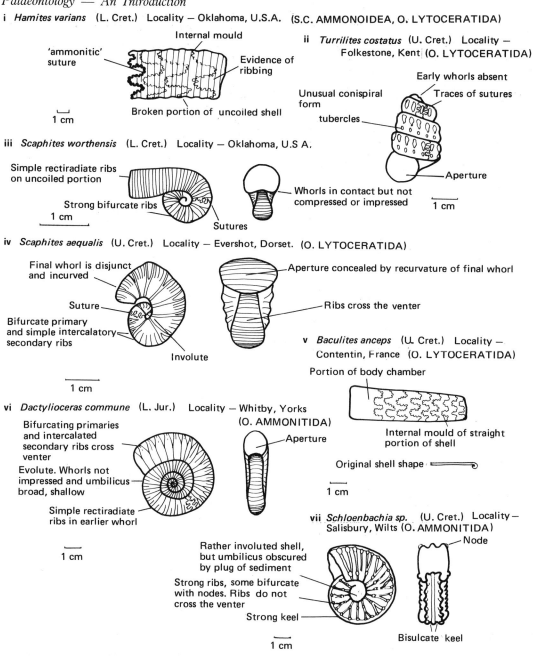

FIGURE 6.17. Lytoceratid and ammonitid ammonoids. (All drawn from original specimens.)

site to settle and metamorphose into the adult. Most bivalve species have separately sexed individuals, but a few are hermaphrodite and (as we have also seen in certain gastropods) some are able to change sex during their lives.

Shell form and soft-part anatomy have all proved to be extremely plastic in the Bivalvia, changing constantly to allow them to adopt many modes of life. In a broad sense, however, bivalves are ecologically specialized animals. The overwhelming majority are filter feeders, and most are burrowers. Some may live at the surface, tethered to rocks or the sediment by a thread-like **byssus**, which in a loosely functional sense, is comparable with the pedicle of the brachiopods. Others cement themselves to firm substrates, while some can actually bore into wood, or even solid rock. There are some (but not many) which have successfully invaded fresh water, and in the Cretaceous an unusual group called the rudists constructed reef masses.

However, structural innovations such as are seen in rudists (see below) are relatively rare. Rather, the success of bivalves has rested upon their ability to form siphons and hence to burrow to a place of safety within the sediment. Together these features have enabled them to exploit the infaunal habitat as no other commonly fossilized group of organisms has done. The different modes of burrowing, as well as the other more unusual life styles, are treated in greater detail in section 6.22 below.

Molluscs 89

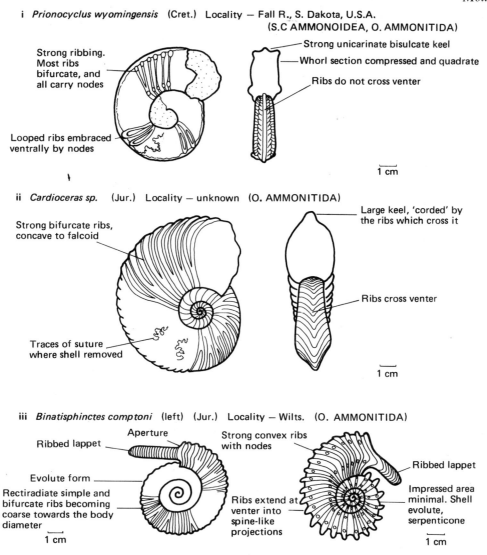

FIGURE 6.18. Ammonitid ammonoids (true 'ammonites'). All drawn from original specimens.

6.21 Shell form

The microstructure of bivalve shells is very variable. On the outside they bear a thin sheath of conchiolin (Fig. 6.19 ii), the same horny, proteinaceous material which composes the ligament. This, of course, rarely survives fossilization. Below this film is a shell of calcium carbonate, but this may take the form of aragonite or calcite, or even both. In the bivalve *Pinna* (Fig. 6.21 f), for example, the outer shell is calcitic while the inner is made of aragonite. And many shells, regardless of mineralogy, often contain a certain amount of conchiolin intermixed within the calcareous layers.

The pattern of crystal growth, which is determined by the mantle, can be very useful when using thin sections and a microscope, but is only very rarely of use when working with hand specimens. And although it has been employed in trying to sort out the complex and obscure taxonomy of bivalves, it too has drawbacks. This is because the different kinds of crystal structure may possess different mechanical properties — some having high tensile strength, others high hardness, and so on.

This means that they may be just as susceptible to adaptational change as, say, shell shape. And in some modern bivalves the very *mineralogy* may be dependent upon temperature, with calcite an increasingly important component among specimens living in colder waters. It is difficult, therefore, to find reliable guides to phylogenetic affinity in the absence of soft parts.

Each valve of a bivalve shell is a curved cone with a very rapid whorl-expansion rate and slight translation (Fig. 6.20). In nearly all cases the two are mirror images of each other, i.e. the plane of symmetry lies along the plane of the **commissure**. This is very different from the brachiopod condition, where the

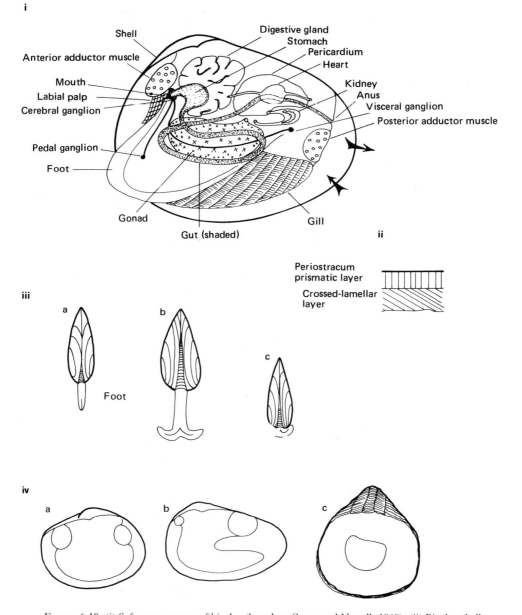

FIGURE 6.19. (i) Soft part anatomy of bivalve (based on Grove and Newell, 1969). (ii) Bivalve shell microstructure. (iii) Operation of the foot during burrowing. (a) extrusion, (b) opening and anchorage, (c) contraction, pulling shell into sediment (modified after Trueman). (iv) Types of bivalve musculature. (a) both adductors of equal size — isomyarian, (b) posterior adductor enlarged — anisomyarian, (c) posterior adductor enlarged, anterior eliminated — monomyarian.

plane of symmetry cuts both valves in two.

Another difference from the brachiopods is the way in which the shell is borne upon the soft parts. Remember that in the archetypal mollusc (Fig. 6.1) the shell is dorsal. This remains so for bivalves, except that in the dorsal region a hinge has developed, and the shell extends ventrally so as to encase the body. Therefore, the **umbones** and the hinge between them are dorsal, and the commissure ventral (Fig. 6.20). Compare this with the brachiopod orientation (Fig. 5.1).

But with no head to mark the anterior, how can we tell which end is 'front' and which 'back'? There is a simple procedure to orientate the shells of bivalves correctly. Its secret lies in the fact that in most bivalves the umbones 'lean' towards the anterior. So, if you hold a shell with the commissural plane vertical and the umbones pointing away from you, then you are facing the posterior of the animal, and the right and left valves are as seen.

This common condition, with the umbones leaning forwards, is called **prosogyrate**. A few forms do exist in which the opposite is the case (e.g. *Nucula* — Fig. 6.22 i) and they are called **opisthogyrate**. With these forms, the procedure described above for valve orientation will not prove correct — but examples are very rare. How would you orientate the valves in a known opisthogyrate specimen?

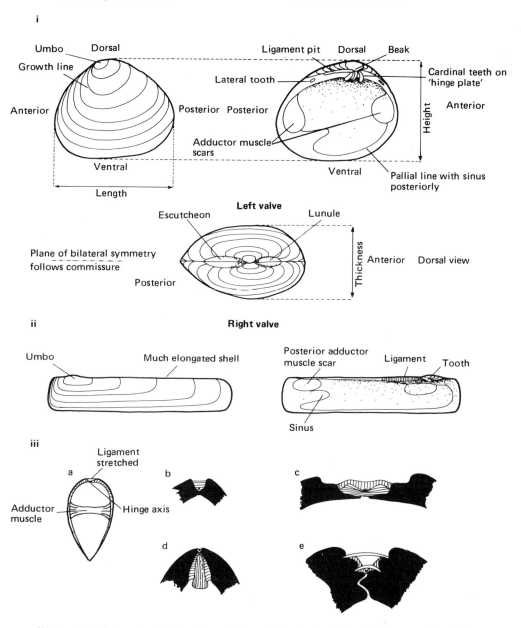

FIGURE 6.20. External and internal morphology of the bivalve shell. (ii) The razor-clam *Ensis* showing extreme elongation of the posterior. (iii) Various types of ligament as seen in transverse section: hinge axis represented by black spot: (a) basic principle of simple ligament under tension in closed shell; (b) simple tensional ligament composed of lamellar tissue; (c) ligament partly tensional, partly compressional i.e. lies on both sides of hinge axis. Fibrous tissue shaded vertically; (d) compressional ligament wholly inside hinge axis; (e) complex ligament, mostly tensional (redrawn after Newell, 1937).

The outer surfaces of the valves invariably bear fine **growth lines** reflecting former positions of the commissure and bearing witness to the incremental nature of their growth. There may also be ornamentation in the form of radial ribs, concentric ridges, and spines.

Turning to the inside surface, we see the impressions made by the soft parts, the dentition on the dorsal margin and the structures associated with the ligament. Just below the umbo, the dorsal edge of the shell is somewhat thickened, forming the **hinge plate**. Projecting calcitic pegs (**teeth**) occur here, together with the **sockets** which locate with the teeth of the opposing valve. Those teeth directly below the umbo are called **cardinal teeth**, those further towards the anterior and posterior extremities of the shell being known as the **lateral teeth**.

In its simplest form, the bivalve ligament (Fig. 6.20 iii) is external, lying on the posterior side of the umbo. It is continuous with the periostracum, but is much thicker and lies in a depression in the shell called the **ligament pit**. It is continuously under tension, pulling the valves open.

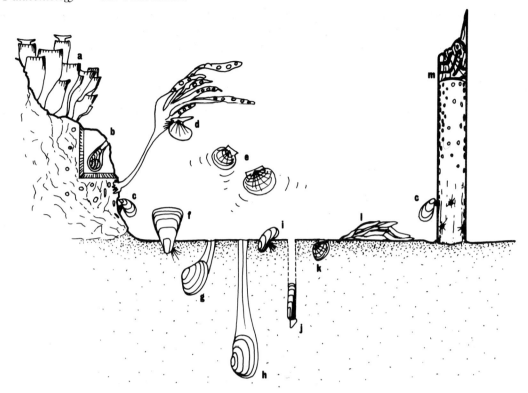

FIGURE 6.21. Modes of life adopted by bivalves. (a) Cemented, coralloid mode of rudists (extinct). (b) Rock-boring as exhibited by *Pholas*. (c) Byssal attachment to hard substrates, e.g. *Mytilus*. (d) Byssal attachment to weed, etc., the 'winged oyster' *Pteria*. (e) Free-lying unattached, capable of limited locomotion usually as an escape reaction as in *Pecten*, the scallop. (f) Semi-infaunal byssate attachment, e.g. *Pinna*. (g) Infaunal burrower with siphons: burrow deep and semi-permanent, e.g. *Mercenaria*, the clam. (h) Infaunal siphonate burrower in a permanent burrow. Soft parts far larger than shell volume — the geoduck *Panope generosa*. (i) A byssate semi-infaunal bivalve liking intertidal sea-grass meadows — *Modiolus*. (j) Infaunal form with short siphons. Razor-clams like *Ensis* and *Solen* can move up and down very rapidly in sediment, burying deeply for protection and rising to the surface to feed. The shell is like a tube when both valves are closed, with permanent posterior (siphonal) and anterior (pedal) gapes. (k) Shallow burrower with short siphons, e.g. the cockle *Cerastoderma* (formerly *Cardium*). Living close to the surface this form is often exhumed and can re-bury itself very quickly. (l) Cemented colonial forms, e.g. the oyster *Ostrea*. Massive shells give stability and protect against borers. Tend to live in monospecific reef-like banks. (m) Wood-boring, as in the 'shipworm' *Teredo*. These almost shell-less bivalves live naturally in driftwood, but since the advent of iron-clad vessels now inconvenience man chiefly by undermining the strength of wooden piles supporting piers, bridges, etc. A recent casualty is the Barmouth railway viaduct, N. Wales, whose replacement-costs as a result of shipworm damage threaten the existence of the Cambrian-coast railway.

But not all ligaments are like this. Some lie in a special area *beneath* the hinge, and instead of being under tension, are under compression, forcing the valves apart like a pencil eraser caught in a door hinge. These are called internal ligaments. Other ligaments may extend *across* the hinge, acting partly by tension, partly by compression.

Perhaps the most notable features of the shell interior are the **muscle scars**. Shells with two adductors are called **dimyarian**, and if the scars are of equal size, they are also called **isomyarian**. If the posterior muscle is bigger than the anterior (and it is always the posterior which is enlarged) then the musculature is **anisomyarian**. Total reduction of the anterior muscle leads to a shell with one muscle only. This commonly migrates to a central position. Such shells, like the oyster (Fig. 6.19 iv, c), are called **monomyarian**.

Around the edge of the shell, paralleling the commissure and apparently joining the two muscle scars in dimyarian shells, is a faint line. It marks the fusion of the mantle to the shell and is called the **pallial line** (remember that another name for 'mantle' is 'pallium' — section 6.2). This line may be indented posteriorly, forming the **pallial sinus**. The presence of this indicates that the animal possessed prominent siphons, for they emerge at this point.

The siphons of the deeper-burrowing kind of bivalve may be very bulky. They may, indeed, be permanently extended and far too large to be fully withdrawn into the protection of the shell. Such species, therefore, have shells which **gape** permanently at the posterior end, even when fully closed. A similar gape may also exist at the anterior end, to accommodate the protrusible foot.

The marred protective efficiency of such a shell is

acceptable to deep-burrowing forms, as it also is to the borers, which live in hard substrates. Borers in rock tend, however, to preserve their shells as rasping organs: wood-borers may reduce them almost entirely, preserving only a pair of vestigial plates to use rather like jaws.

Although, as has been said, most bivalves are symmetrical about their commissural plane, those which live reclined on the sea floor or cemented to rocks tend to develop larger lower valves with smaller, lid-like upper valves. Such forms, like the *Pecten* or *Ostrea*, are termed **inequivalve**.

You will find examples of bivalves displaying all the features here described in Figs. 6.22–6.26. From the foregoing discussion, you will have gathered that the form of bivalves is very closely related to their mode of life. It is for this reason that the major part of what follows discusses this relationship in more detail, to enable you to make full use of the ecological information which bivalves can give to the geologist.

6.22 Mode of life

Three basic modes of life are exhibited by bivalves: burrowing and boring, byssal tethering and cementation. To these might be added a minor group of free-lying types, some of which may have limited swimming ability. These are all illustrated in Fig. 6.21.

6.22.1 Burrowing and boring

Burrowing may be very shallow, as exhibited by the cockle *Cerastoderma*, or deep, as exemplified by *Mya*

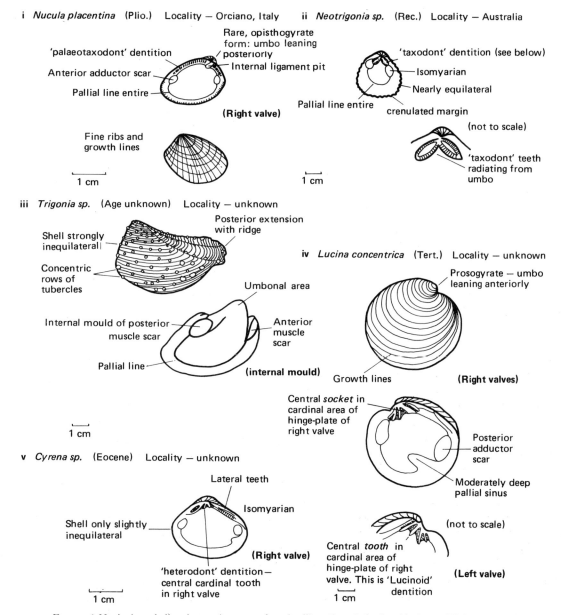

FIGURE 6.22. Active, shallow-burrowing or surface-dwelling, deposit-feeding bivalves. (All from original specimens.)

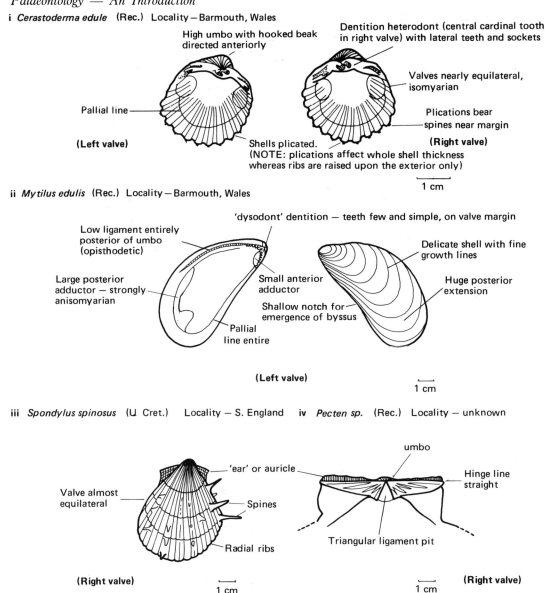

FIGURE 6.23. Shallow-burrowing (*top*) and sessile, byssate bivalves (ii), (iii). Hinge area of free-lying, active bivalve (iv). (All from original specimens.)

and the razor clams *Ensis* and *Solen* (Figs. 6.23 i, 6.26 i and 6.20 ii).

Cerastoderma has a thick shell for its small size. It is equivalve, has a rounded outline and is adorned with radial ribs. These are created by a folded commissural plane whose crenellations intermesh like the teeth of a gear wheel when the shell is closed. The shell has no permanent gape. Internally there are two isomyarian scars and a pallial line which has no sinus and is said to be **entire**.

This is a shell built for protection. It can close completely on the soft parts and its meshed commissure resists the battering of waves and the attacks of predators by preventing easy dislocation of the valves. *Cerastoderma* lives just below the sediment surface in vigorous tidal environments. It therefore needs only short siphons — hence the lack of a pallial sinus — and is capable of reburying itself rapidly if exhumed.

The outline of the cockle shell is rounded. Generally, the deeper a bivalve buries itself, the more posteriorly enlarged its shell becomes. The shell of *Mya* is a good example of this. The protection afforded by the burrow (which can be 50 cm deep) means that the shell can be much thinner than that of the cockle. Its valves need not mesh, and indeed they do not even close properly; wide siphonal and pedal gapes are both well developed. The long siphons have created a deep pallial sinus, and they are so large that they need to lie permanently outside the shell. Dentition is weak. In Fig. 6.26 i you will also see the distinctive projection which bears the internal ligament. It is called the **chondrophore**.

Shell elongation is carried to extremes in the razor clams, whose valves form an open-ended tube. Although they are deep burrowers, their siphons are fairly short, so that the animal must rise in its long

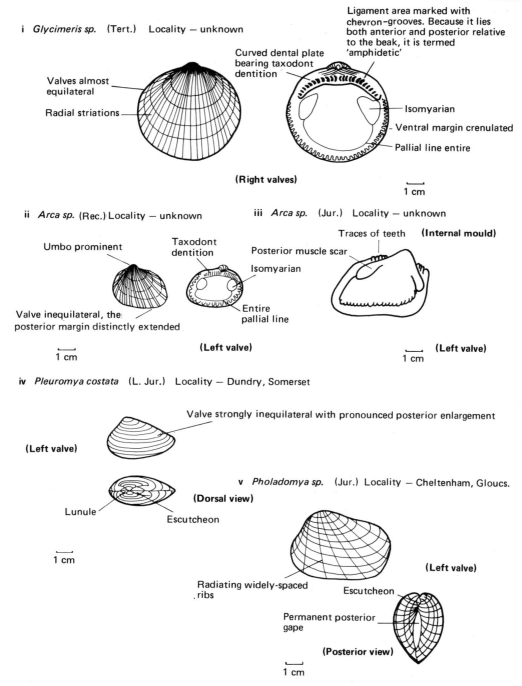

FIGURE 6.24. Semi-infaunal and burrowing bivalves. (All from original specimens.)

burrow to feed. It can, however, retreat downwards very quickly if disturbed. Once again, the dentition is weak and the shell thin.

How do bivalves burrow? In its simplest form (Fig. 6.2 iii) the cycle begins with the opening of the valves. This forces them against the burrow walls, locking the shell in position and providing a firm anchor against which the foot can push as it forces downwards. When fully extended, the foot swells at its end; the valves close and water in the mantle cavity is expelled, liquefying the sediment around the shell. The foot then contracts, pulling the shell down through the liquefied sand. The cycle is repeated until the required depth is attained.

Bivalves which bore into rock or wood carry the reduction of the shell's protective role even further. The vestigial shell of the shipworm *Teredo* has been mentioned. Rock-borers have permanently gaping shells, and *Pholas* (Fig. 6.26 ii) also boasts rasp-like ornament which enhances the shell's abrasive properties.

The use of the shell as a reamer requires a powerful and unconventional musculature. Extra attachment area is provided by the projecting **myophore** (Fig.

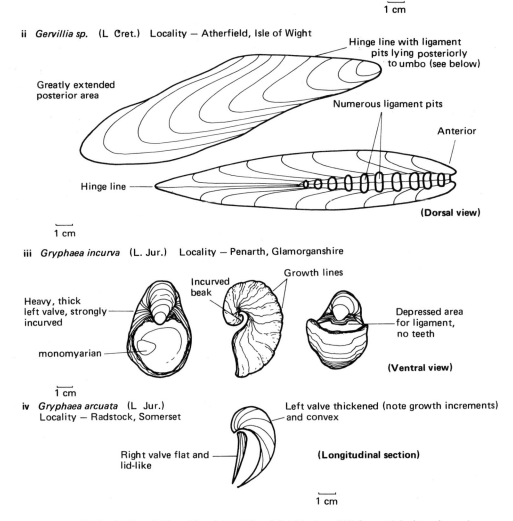

Figure 6.25. Sessile (i) and (ii) and free-lying (iii) and (iv) bivalves. (All from original specimens.)

6.26 ii), and, since a ligament would not provide enough opening force for the movement to have any real abrasive power, one of the adductor muscles has come to attach itself to the *exterior* of the shell at one end, so becoming a 'diductor' muscle. Its emplacement is protected by the secretion of **accessory shell** in the umbonal region.

Another rock-borer, *Lithophaga* (literally, 'rock-eater'), in addition to mechanical abrasion, can produce acidic mucus to soften the substrate and ease penetration. Its proteinaceous periostracum protects it from dissolving its own shell. In a similar way, *Teredo* can soften wood by producing secretions, but these are actually digestive, and thus provide the animal with additional nutrition.

6.22.2 Byssal tethering and free-living forms

The byssus, a skein of horny threads made of conchiolin, is secreted by a gland near the base of the foot. Each thread begins as a liquid extrusion of this gland, but soon becomes tough and fibrous on contact with water. The byssus may serve to attach the bivalve to a rock or to seaweed (known as the **epibyssate** habit) or may anchor the shell in sediment, rather like a root system (the **endobyssate** habit).

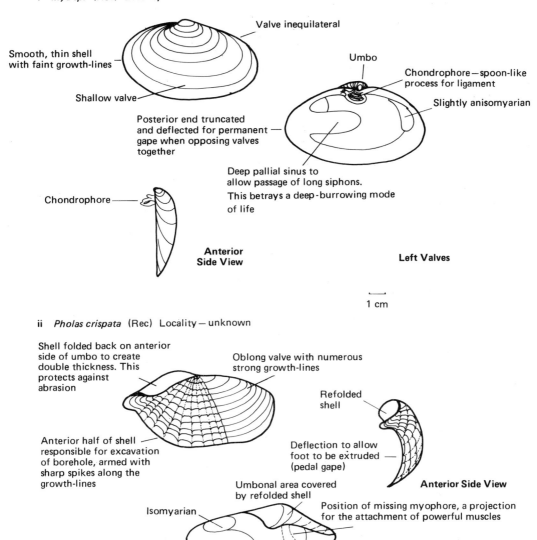

FIGURE 6.26. A deep-burrowing and a boring bivalve. (All from original specimens.)

In most byssate forms the commissural plane is held vertically. In the mussel *Mytilus* (Fig. 6.23 ii), the shell's anterior portion is much reduced, and the posterior projects upwards from the substrate, so allowing for very close packing of individuals in clusters. This is a protective device, making mechanical dislodgement by predators (such as the seagull) more difficult. The byssus emerges from a shallow notch in the antero-ventral area of the commissure. In other forms this **byssal notch** may be very deep.

Byssal tethering is seen in many bivalves in their early, post-larval stages. In burrowers and free-lying forms it is lost fairly quickly. Forms whose adult form is byssate therefore preserve what is a generally youthful feature in very many bivalves. Such retention of a youthful characteristic in adulthood is known as **neoteny**.

The free-lying scallop *Pecten* soon loses its byssus. In adulthood it can swim to escape danger by clapping its valves together. Water may be expelled at various points around the edge — wherever the mantle is commanded to form a temporary outlet. This means that considerable manoeuvrability is possible. Usually water is expelled in the hinge areas, so that the scallop appears to swim by 'taking bites' out of the water ahead of it.

We have looked at free-lying brachiopods (Ch. 5) which developed shells that, by virtue of their shape, tended to be self-righting. A similar set of devices, conveying extreme stability in one position, is seen in free-lying bivalves.

Weight is an obvious stabilizing influence, and the enlargement of the lower valve coupled with thickening of the shell is frequently seen in free-lying species.

Others may extend one or both ends of the hinge line into long fingers, recalling the extended hinge lines of spiriferid brachiopods. These are thought to act, like the outriggers of a boat, to prevent overturning.

Gryphaea (Fig. 6.25 iii, iv) has an enlarged and incurved lower valve. This gives the animal a horn-like shape. This is a shape seen in many soft-bottom recliners; solitary corals, sponges, aberrant free-living barnacles and many others all display it. The reason for its evolution in all these unrelated groups is its functional efficiency as a self-righting form. It is a shape which lies naturally with the aperture pointing upwards. If disturbed from this attitude, currents in the water induce scour around it in such a way that it will gradually sink back into its original orientation.

The largest of all bivalves, the giant clam *Tridacna*, has a strongly plicated shell. Some species are byssate, others free-lying, but their most remarkable feature (apart from their size) is the fact that the wide outer edges of their fleshy mantles carry symbiotic algae. Although not a unique feature among bivalves (the primitive protobranch *Solemya* carries symbiotic bacteria), it is one most usually associated with scleractinian corals (Ch. 8).

Free-living forms frequently developed from byssate types by simple atrophy of the byssus. It is likely that cementation came about through byssal mineralization.

6.22.3 Cementation

The larva of the oyster settles on a substrate and is fixed, in the first instance, by a sticky secretion of the byssal gland. Thereafter it is the mantle edge of the lower valve which cements the shell. In oysters it is the left valve which becomes fixed in this way. Cementing groups vary in the valve by which they choose to become affixed; rudists (see below) chose the right.

Oysters are among the most successful bivalves. Each is able to filter about a barrel of water per day, and they often grow in such profusion as to form thick beds consisting of little else but oyster shells. In the North Sea, south of the Dogger Bank, an area of several hundred square miles is completely clothed in oysters.

Oysters are good indicators of shallow seas, and may be well developed along shorelines, a factor which is of help in palaeogeographic reconstruction. Like free-living forms, the shells are massive, but for not entirely similar reasons. For although stability is no handicap, rapid growth is important when competition for space is high. Also, a thick shell helps ensure that parasitic shell-borers are not able to pierce the protective armour.

In cemented forms the need to conform to the substrate and the close pressing of other individuals makes shell form very variable. Competition for space is usually a prominent factor of life for encrusters, and in extreme circumstances leads to upward growth, just as the same pressure of space leads to the building of skyscrapers in restricted areas like Manhattan.

The rudists were very successful Cretaceous bivalves, whose fixed valves grew upwards like organ pipes, often to heights of over half a metre. Many were smaller, but very large individuals formed reefs in the Tethyan Realm (Southern Europe, the Mediterranean and the Middle East) at this time.

The animal itself, however, remained quite small, supported on its immense, columnar shell. The lid-like left valve was modified so as to move vertically up and down rather than rotate about a hinge, and bore a single elongate tooth which meshed with two in the fixed valve. Some were perforated like gratings, through which the inhalent and exhalent currents apparently passed.

6.23 Geological history and stratigraphic value

Bivalves are thought to have sprung from a minor, extinct molluscan class, the Rostroconchia. These, appearing in the L. Cambrian, had a shell formed of two valves — though these were not hinged. They appear to have been rigidly (or semi-rigidly) fixed together, but the freeing of this dorsal area may have set off the evolution of the true bivalves.

The first bivalves also appear in the L. Cambrian, the earliest example being *Fordilla*. Once again, however, we see that it was not until the L. Ordovician that the great expansion of the class began, with the development of the protobranch deposit feeders like *Nucula* (Fig. 6.22 i). Deep burrowing also began then, as well as byssal attachment — although the burrowers of that time were 'lucinoid' bivalves, which have no true inhalent siphon. They use a mucus-lined tube instead.

In the early Mesozoic, actively burrowing siphonate forms gave rise to the second great radiation. The addition of cementing and boring forms also contributed to the bivalves' present pre-eminence, though there is some evidence that the boring habit may have originated much earlier, possibly in the Ordovician.

By and large, bivalves are not very useful stratigraphically, though the rapidly evolving rudists are used to zone the Tethyan Cretaceous. Moreover, non-marine bivalves of the Coal Measures (U. Carboniferous) have been of great stratigraphic and economic value. Marker bands based on *Carbonicola*, *Anthracosia*, *Anthraconauta* and *Anthraconaia* (among others) were set up earlier this century by Sir Arthur Trueman. This enabled mining engineers to identify their seams with confidence and to predict the occurrence of other coals deeper in the sequence.

7. Echinoderms

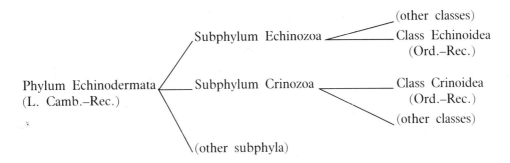

7.1 Introduction

Echinoderms are exclusively marine and include the sea urchins (Echinoidea), sea lilies (Crinoidea), starfish (Asteroidea), brittle stars (Ophiuroidea), sea cucumbers (Holothuroidea) and extinct classes Cystoidea and Blastoidea. In this book, however, we shall only be concerned with the echinoids and crinoids, which are by far the most significant groups palaeontologically.

Echinoderms as a whole are unique for a number of reasons. To begin with, they may be among the closest invertebrate relatives of the Phylum Chordata. Secondly, they have internal skeletons, but unlike ours, echinoderm skeletons are made of plates embedded so shallowly in the outer body layers that in many respects they are functionally external — that is, they mostly enclose the soft parts, rather than supporting them from within.

Thirdly, echinoderms have a distinctive body plan based, in most cases, upon a five-rayed or **pentameral** pattern. Depending upon the angles between these rays, the plan may be radially symmetrical (angles all equal) or bilaterally symmetrical (angles not all equal). In turn, the choice of radial versus bilateral symmetry seems to depend largely upon mode of life.

The fourth distinctive feature is the **water vascular system**. This consists of a network of internal plumbing, filled with sea water, which is capable (by hydraulic action) of extruding muscular **tube-feet** from holes in the skeleton. These feet have many functions, including locomotion, respiration and food gathering. Although they only allow sluggish movement in mobile forms, they can also enable some limited predation; the starfish, for example, can prise open bivalve shells by the exertion of sustained tension.

Echinoderms are dominantly calcitic, and so tend to fossilize well. But their skeletal construction means that after death they often disintegrate. Echinoderm plates and spines are important sediment-forming materials as a result of this, and are especially abundant in certain limestones.

I. ECHINOZOA — CLASS ECHINOIDEA

7.2 Shell form and soft parts of a regular echinoid

An echinoid skeleton (or **test**, as it is called — Fig. 7.1) can be viewed as a semi-rigid sac containing more-or-less fluid soft-parts, and as such the form of a typical regular echinoid tends strongly toward that of a rubber balloon filled with water and allowed to stand upon a flat surface. The test itself is built up of closely fitting plates of porous calcite, each plate being crystallographically uniform.

The terms 'dorsal' and 'ventral' are not used when speaking of these animals. Instead we talk of positions relative to the mouth. The side where the mouth lies is referred to as the **oral** side, and the side opposite to it is called **aboral**. In regular echinoids the mouth is on the underside.

Lying centrally in this aboral surface is the **apical disc** (Fig. 7.1 iv), at the convergence of the radial pattern. Here a double ring of plates encircles the **periproct**, which is the anus of the living animal. The inner ring is composed of larger, **genital plates**, so called because a **genital pore** opens upon each of them. From these pores the sperm of male echinoids is shed into the water to fertilize the eggs, which escape from similar pores on the female. All bar one of these genital plates are of the same size. The one larger plate is obviously very porous, its many perforations allowing the water vascular system to communicate with the exterior. This is the **porous** or **madreporite plate.**

Outside the genital ring lies a ring of **ocular plates,** each with an **ocular pore.** This pore is associated with the water vascular system, and is discussed below.

The test may be divided into ten radial segments extending between the apical disc and the mouth, which lies centrally on the oral surface (Fig. 7.1 ii). Five similar, narrower sections contain the tube feet in the living animal, and they join aborally with the ocular plates of the apical disc. They are called **ambulacra** (or **ambs**), and between each ambulacrum and the next lies a wide **interabulacrum** (**interamb**) which terminates aborally at a genital plate.

Interamb plates are imperforate, but often bear knobs or tubercles (Fig. 7.1 v). These are the articulation bases for the many spines which bristle from the test surface in life. Spines are rarely preserved in place, but occur commonly in bioclastic limestones. They tend to be longest at the equatorial region of the test, the area of greatest girth which is known as the **ambitus** (Figs. 7.1 iii, 7.2, 7.3). They are used for locomotion on sandy surfaces where the tube feet cannot gain purchase, but they can also help to wedge the animal into crevices. They have considerable protective function, being often sharp and brittle — even irritant if allowed to penetrate the flesh. On a

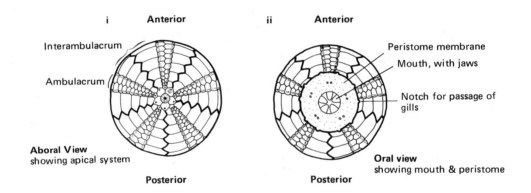

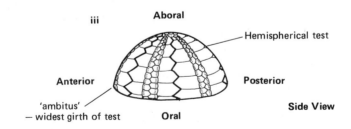

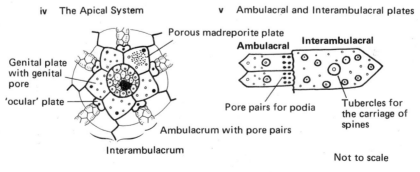

FIGURE 7.1. Skeletal morphology of *Echinus* — a regular echinoid.

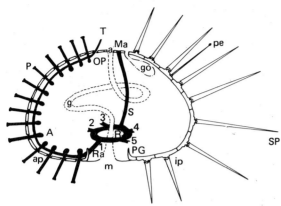

Key:-
Ma — Madreporite or porous plate leading to water-vascular system
R — Ring canal or circum-oral canal
Ra — Radial canal (five in all)
A — Ampulla
P — Podia or tube-feet
S — Stone canal, connecting the madreporite to the ring canal
T — Tentacle emerging through single pore on ocular plate
PG — Perignathic girdle
SP — Spines
OP — Ocular Plate
m — mouth
a — anus
g — gut
go — gonad
pe — pedicellaria
ap — ambulacral plate
ip — interambulacral plate with tubercles

FIGURE 7.2. Transverse section of a regular echinoid. (Modified after Moore, Lalicker and Fischer.)

smaller scale, the test surface is protected by **pedicellariae**, minute pincers sometimes invested with poison glands (Fig. 7.3 iv).

Both ambs and interambs are composed of two rows of plates meeting centrally at a zig-zag suture (Fig. 7.1 v). Ambulacral plates bear **pore pairs** near their outer edges, each pair in life leading to a single **tube-foot** (see below). Both ambs and interambs widen to the ambitus and narrow again on the oral surface, where they terminate at the **peristome**. This is a large hole where the mouth and the complex jaw apparatus known as **Aristotle's lantern** once sat.

Interambs meet the edge of the peristome at small notches where respiratory gills protrude from the test to supplement the tube feet in gaseous exchange. From the mouth the gut extends upwards through the spacious body cavity towards the anus (Fig. 7.2).

Close to the anus is the madreporite plate. It communicates with an internal, vertical tube composed of calcified rings, called the **stone canal** (Fig. 7.2). This links up with a **circum-oral canal** from which five **radial vessels** extend, each one passing up the centre of an ambulacrum and terminating in a blind-ending tube which protrudes through the single pore in the ocular plate.

All along these radial vessels, paired tubes branch off regularly. Each leads to the base of a tube foot where there is a muscular sac or **ampulla**. Two tubes leave this sac to pass through a pore pair and lead into a single tube-foot (Fig. 7.4 ii).

Contraction of the ampulla forces water into the tube-foot to extend it. The tube-foot has a sucker at its end by which to cling to the substrate. Upon contraction of the ampulla, the foot is withdrawn and the animal is dragged along, by the concerted action of many hundreds of such tiny podia.

7.3 Irregular echinoids

In irregular echinoids the test is bilaterally symmetrical and takes on a somewhat heart-shaped, rather than circular outline (Fig. 7.5 i). It is also somewhat flattened, even assuming biscuit-like form (Fig. 7.5 ii).

But there are many other differences from the regular echinoid type. The spines, for example, are short and, instead of sticking out aggressively, are plastered down over the test like coarse fur. Also, the anus is no longer within the apical cluster, but lies some way away — even migrating onto the oral surface (Fig. 7.5 ii). Moreover, the peristome is much smaller, reflecting a much lighter system of jaws, and is usually not in the centre of the oral surface (Fig. 7.5 i).

On the aboral side, the pattern of ambulacra is quite different. One amb, lying in the plane of symmetry, has become much enlarged to form a groove which runs down, across the ambitus — so forming the notch in the heart-shaped outline. The four other ambs are expanded aborally into four broadly depressed areas called **petals**, which terminate above the ambitus. Below this they are of normal type (Fig. 7.5). The variation in amb pattern is an expression of the differentiation of tube-foot function (see below).

On the oral surface, and expanding back from the lip which overhangs the peristome, is a raised, tuberculate area called the **plastron** (Fig. 7.5 i) which in life is strongly spinose.

7.4 Mode of life

Bilaterally symmetrical echinoids are so shaped as an adaptation to burrowing. But how, exactly, does it help them? Let us look first at the mode of life of regular types.

The common, pinkish tests so often on sale at British resorts and used as ornamental lampshades in the lounges of seaside boarding houses, are of the species *Echinus esculentis* (Fig. 7.1). It is a spinose creature living mostly on rocky sea floors and ranging down to as much as 1000 m. Its tube-feet can cling so securely to rocks that even surf cannot dislodge it, and with its massive lantern it browses upon seaweed and devours encrusting flora and fauna. On sand it moves by walking upon the spines of the oral surface.

In other regular echinoids the jaws can work together with the spines to excavate boreholes in rock,

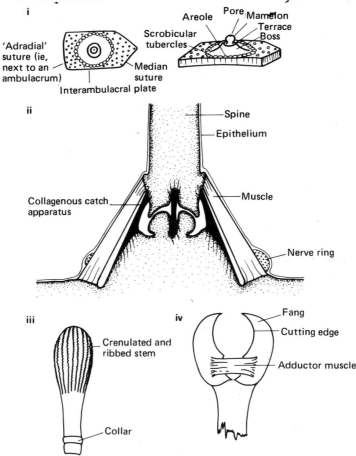

FIGURE 7.3. Echinoid appendages and their attachment. (i) tubercle morphology; (ii) cross-section of spine, tubercle and soft tissues (redrawn after A. B. Smith, 1984); (iii) a club-like cidaroid spine; (iv) a pedicellaria.

and this activity combined with the general activity of browsing types can be a major source of 'bioerosion' in reefs. Some such crevice-dwellers expand their holes as they grow and produce bottle-shaped burrows from which they are unable ever to escape. Such behaviour is common in heavily wave-battered environments.

Irregular echinoids have taken on all the problems of burrowing organisms: namely, how to breathe, eat and excrete without breathing in what has been breathed out or eating that which has been excreted. It is in ensuring that these things do not happen that their altered shape conveys its advantage.

The heart urchin *Echinocardium* lives about 20 cm below the sea bed, connected to it by a single tube in the sand, maintained by much-extended tube-feet with bristly terminations like a chimney-sweep's brush. This is the incurrent funnel (Fig. 7.4 iii). A similar, but blind-ending, tube extends from the anal area, and this serves as a sanitary 'soakaway'.

The inhalent chimney ends above the apical portion of the enlarged anterior amb. This is surrounded by a smooth region (or **fasciole**) where ciliated epidermis and small, ciliated spines create the downward current. This passes over the anterior amb, where food-gathering tube-feet capture food particles and convey them to the mouth.

Other currents, dividing at the foot of the chimney from the main flow, sweep over the petals where the tube feet have a respiratory function. Finally, currents created in a similar way by the sub-anal fasciole carry wastes away down the drain. When this is full, *Echinocardium* moves off, retracting its specialized tube-feet and moving anteriorly by the rowing action of the flattened spines of the plastron and by the digging action of those around the anterior amb itself.

So *Echinocardium* is a filter feeder and has no need of a heavy jaw apparatus. The loss of the complex 'lantern' thus permits the mouth to become reduced in size.

7.5 Spired echinoids, sand dollars and echinoid infaunalism

Echinocardium and genera such as *Micraster* (see also Ch. 12) display the straightforward adaptations seen in very many burrowing echinoderms. The origin of irregular form was in itself a remarkable evolutionary event, which involved many drastic modifications to

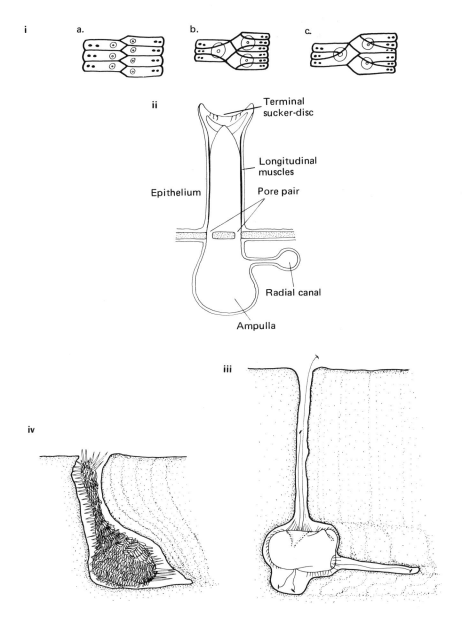

FIGURE 7.4. Ambulacral plates, tube-feet and life-habits. (i) Types of compound plate. ia — simple ambulacral plates, each an independent unit. ib — 'diademoid' compound plates; three simple plates fuse together into one. The central member is largest of the three. ic — 'echinoid' compound plates. Three units per plate, the lowest member being the largest. Note how the pore-pairs remain faithful to their original plate, but that tubercles may be formed which embrace more than one simple ambulacral plate and relate to the compound unit. Compound plates are typical of most regular echinoids of Mesozoic and Caenozoic. (ii) Cross-section of a tube-foot showing how it relates to the pore-pair on the ambulacral plate. There is a one-way valve between the radial canal and the ampulla which closes when the ampulla contracts to extend the tube-foot. The valve prevents return of water into the radial canal. (iii) Mode of life of the modern heart-urchin *Echinocardium cordatum*; see text for explanation (after Nichols, 1959). (iv) Reconstruction of the Cretaceous echinoid *Hagenowia blackmorei* showing the characteristic rostrum which served to connect the animal with the surface in the absence of specialized tube-feet. See text for full explanation (after Gale & Smith, 1982).

Note in (iii) & (iv) the sedimentary traces produced by the movement of these infaunal echinoids through the substrate. Movement in both cases is from right to left and took place episodically, probably when the sanitary drain became full to capacity.

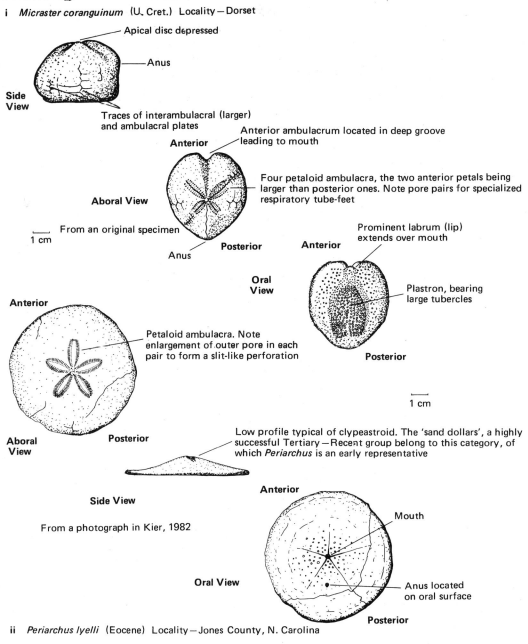

FIGURE 7.5. Cretaceous and Tertiary echinoids.

both test and soft parts, yet it was completed in the 10 million years between the Sinemurian and Toarcian (L. Jurassic). Still more spectacular evolutionary leaps were in store for infaunal echinoids, however.

During the Cretaceous the infaunal habit became established. Yet there were some very unusual forms around at that time which lacked the specialized chimney-brush tube-feet and so instead developed a **rostrum** (Fig. 7.4 iv). This was a snorkel-like proboscis which extended to the surface, so allowing the animal to lie deep while its slender construction minimized the chances of detection by predators.

During its evolution, *Hagenowia* also reduced its general dimensions. This is thought to have been an adaptation to improve gaseous exchange; the size reduction would have greatly increased the surface-to-volume ratio (section 10.3). Other burrowers, however, increased respiratory efficiency by enlarging the petals, thus raising the absorptive area by the creation of more specialized respiratory tube-feet. This tactic was well exploited by the clypeastroids (Fig. 7.8 i).

These appeared in the Palaeocene, and from their stock evolved the most successful modern group of echinoids, the sand dollars (Fig. 7.5 ii). Theirs was a meteoric success, from inception in the early Eocene to worldwide distribution by Middle Eocene time.

Sand dollars possess many structural adaptations related to shallow burrowing in sand under frequently very rough conditions. Their flattened tests ease movement through the sediment; their large petals

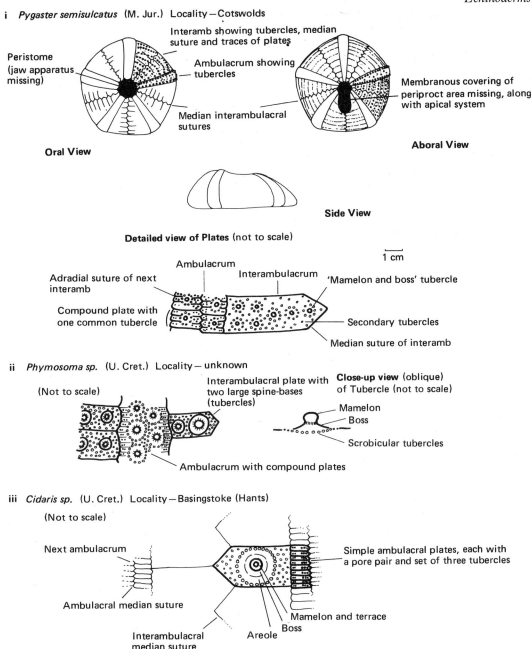

FIGURE 7.6. Jurassic and Cretaceous echinoids. (All from original specimens.)

enhance respiration and their unique internal pillars and walls strengthen the test considerably. Their plates have a more stress-efficient microstructure, and they are linked firmly to one another by collagen fibres.

To improve feeding efficiency, sand dollars diversified their ambulacral canals and developed extra **accessory** tube-feet. These are served by **microcanals** which actually permeate the test plates, and are best seen using X-rays. Other canals, known as **macrocanals** and formed from the internal walls mentioned above, are used by the juveniles of certain species to store grains of magnetite. These they select from the sand and use to increase their body mass by up to 23%. By so doing, these juveniles increase their stability, though the habit has effectively restricted these species to beaches where such grains occur.

Such modifications have made sand-dollar tests very resistant and increased their fossilization-potential considerably. But all burrowing forms have a much greater chance of preservation, and indeed, their fossil record is much less patchy than that of the epifaunal, regular echinoids.

7.6 Geological history and stratigraphic value

From their origins in the Ordovician, echinoids made steady but unremarkable progress through to

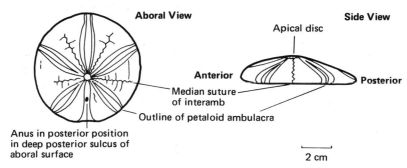
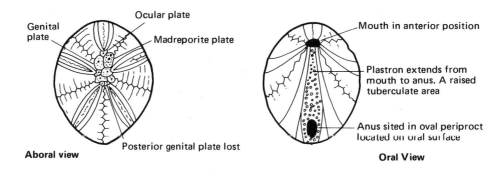
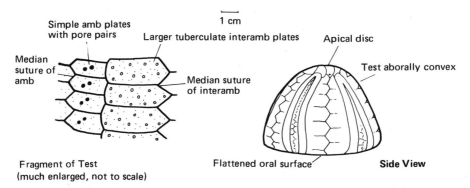

FIGURE 7.7. Jurassic and U. Cretaceous echinoids. (All from original specimens.)

the Carboniferous, by which time their forms were diverse but their numbers not great. The end-Permian extinctions all but wiped them out, but the class survived through one, lone genus, *Miocidaris*. From this single ancestral genus all post-Permian echinoids derived. There were many such descendants, and since the Triassic echinoids have remained significant components of the marine fauna as opportunities offered by new modes of life were taken up and successfully exploited.

Echinoids are stenohaline (section 5.5), and as such they are very reliable indicators of the marine conditions they favour. In between episodes of startlingly rapid evolutionary development, they have shown gradual changes in morphology which have allowed them to be used as zone fossils, especially in the Chalk. The use of *Micraster* species in the Cretaceous (Ch. 12) is perhaps the most famous example, but clypeastroids are also much used in Tertiary rocks.

II. CRINOZOA — CLASS CRINOIDEA

7.7 Skeletal construction

The crinoids are typically fixed, stalked creatures which grow either in dense groves in shallow water or rather less abundantly in water of great depth. The **stem** rises from the sea bed (where it may be attached by root-like appendages) and attaches to a cup (**calyx**) which is equivalent to the body in other echinoderms. Unlike most of them, however, it is the oral surface which is directed upwards, and the rest of the body is drawn out into **arms** (Fig. 7.8).

The stem is composed of columnar plates called **columnals** or **ossicles**, which may be discoidal or stellate in plan and which articulate freely together. They also have a central hole down which the soft parts extend from the calyx. The calyx plus its arms are collectively known as the **crown**.

At the top of the stem the ossicles meet the lowermost plates of the crown. These are five in number and are called **basals**. This holds true for all forms except those in which three extra plates, the **infrabasals**, interpose between stem and basals. Lying above the basals are five **radials**. These articulate with the lowermost plates of the arms, which are called **brachials**. The brachials may be free or may be incorporated in the calyx in some forms.

Forms with only radials and basals are called **monocyclic**. **Dicyclic** crinoids are those with infrabasals. The scheme described here is the basic one upon which calyces are constructed. But many complications can arise which may obscure the fundamental plan, such as **compounding** (transverse subdivision) of radial plates into a lower cycle of **infraradials** and a higher one of **supraradials**.

The arms subdivide — often many times. They have a grove on their upper surfaces where a ciliated **food groove** runs, taking food towards the mouth. These grooves are continued on the oral surface of the calyx by five ambulacra. This oral surface is typically membranous, but it may contain plates, and at its centre the mouth is to be found. The anus is usually situated in one of the five interambs, but it is often raised upon an **anal tube**. There is no madreporite.

7.8 Mode of life

Although crinoids are still alive in modern seas,

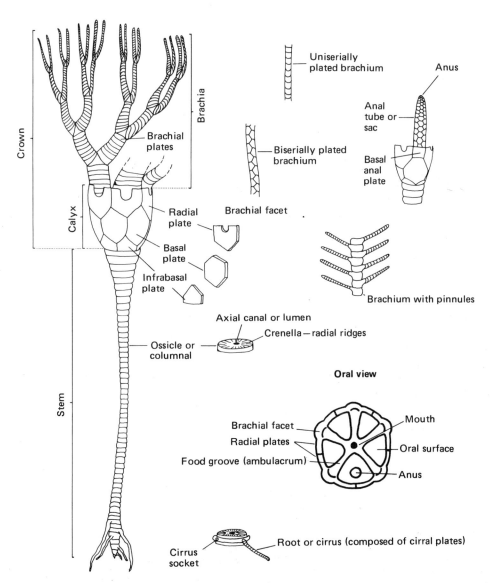

FIGURE 7:8. Basic crinoid morphology (no scale).

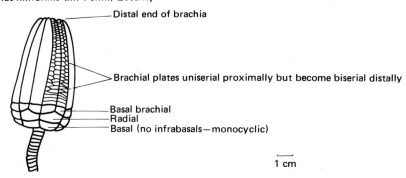
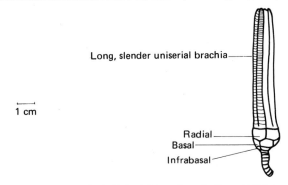
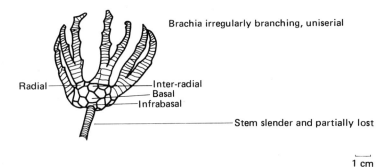

FIGURE 7.9. Carboniferous and Permian crinoids. (All from artificial casts.)

they seem to have undergone a major change of habit in the recent geological past. That is, they have either moved from shallow to deep-water habitats (e.g. *Apiocrinites*, Fig. 7.10 ii) or they have ceased to be attached, lost their stalks and taken to an active life. Today these unattached forms are the more common, living at all depths and emerging from crevices at night to perch upon rock or weed and face into prevailing currents to filter food particles in their feathery arms. Others, living where there are no currents, merely spread themselves out flat and feed on whatever happens to settle upon them.

By analogy with these two feeding habits, it is asssumed that crinoids have always had these two life styles open to them. We must therefore look to skeletal morphology to try to distinguish which forms followed which mode of life.

Crinoids which faced into currents will have had stems flexible near the crown, and to aid in filtration of the current the arms would need to be feathery, i.e. **bipinnate** (Fig. 7.8), bearing two rows of **pinnules**. Those which could not have bent their cups through 90° so as to face currents may well have adopted the other, more passive, life style. Their stalks are less likely to have attained great length, and their arms less likely to have been pinnate.

7.9 Geological history and stratigraphic value

In the Palaeozoic, crinoids were by far the most abundant echinoderms. They appeared in the L. Ordovician, became important sediment builders in the shelf seas of the Silurian and maintained this position in the succeeding Devonian and Carboniferous — occurring in dense accumulations and building great thicknesses of crinoidal limestone.

Echinoderms 109

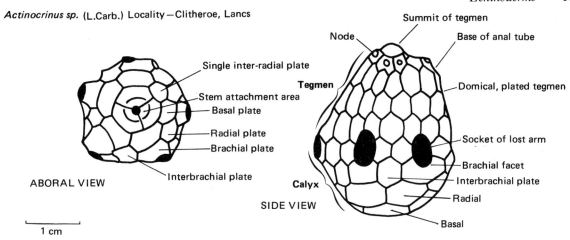

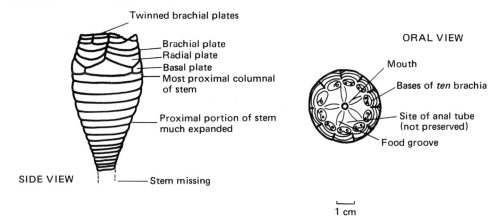

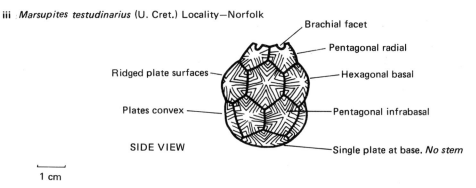

Figure 7.10. Assorted crinoid calyces ((i) & (iii) from original specimens, (ii) from an artificial cast).

At the end of the Palaeozoic, however, the three existing orders of crinoids died out in the extinctions which marked the end of the Permian. Mesozoic and modern crinoids all belong to a different order which, since Cretaceous times, has seen the increasing dominance of stemless forms.

Crinoids have too patchy a distribution on sea floors to realize the potential for correlation which their fairly rapid patterns of evolution might otherwise have afforded. They are therefore somewhat rarely used, but can give excellent results in certain formations. In S.E. England, for example, *Marsupites* (Fig. 7.10 iii) is a zonal index fossil for the U. Cretaceous.

8.
Corals and Stromatoporoids

I. CORALS

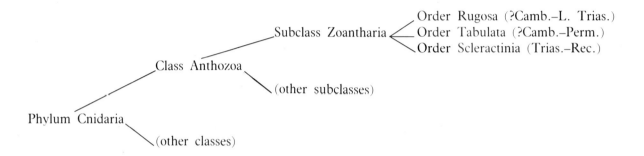

8.1 The coral animal

The individual coral animal or **polyp** is a simple, jelly-like organism consisting of a sac-like body with a mouth at the top, surrounded by tentacles (Fig. 8.1 i). It is constructed by an inner, digestive **endoderm** and an outer **ectoderm** which is sensory and bears stinging cells for protection and capture of prey. The whole animal is contractile and can retreat if disturbed. Between the ectoderm and endoderm lies a gelatinous layer, non-cellular in composition, which is called the **mesogloea**. It is the elasticity of this layer which restores the polyp to its proper shape once the animal relaxes again after retracting.

The body cavity or **coelenteron** is divided by radially-emplaced curtains of tissue called **mesenteries**. These serve to increase digestive and absorptive area. Digestion takes place not actually within the coelenteron but inside the cells which line it. These engulf the food once it enters the body, and digest it intracellularly within **vacuoles**. Waste products are removed via the mouth. Different orders of zoantharia (corals) have different numbers of mesenteries.

Reproduction is either asexual (budding of new polyps by the parent) or sexual (release of male and female gametes into the water). The budding process turns a solitary polyp into a compound individual, the continuation of the process resulting in a mass of polyps called a **colony**.

8.2 The coral skeleton

The coral polyp secretes a calcareous skeleton or **corallum** (Fig. 8.1 i). This takes the form of a cup (**calyx** or **calice**) upon which the animal sits. There is a **basal plate**, and radial walls (**septa**) extend from the periphery towards the centre. These septa lie in between the mesenteries and so reflect their number. Since, as has been said above, the different coral orders have distinctive numbers of such mesenteries, the septa are of great taxonomic importance to the palaeontologist.

At later stages of development, as the polyp grows in size and the cup enlarges, more mesenteries are created and so more septa need to be inserted. The first-formed septa (**prosepta**) are usually recognizable, being generally longer and thicker than the secondary **metasepta**. All septa join the peripheral wall of the calyx, which is called the **epitheca.**

The material of which the skeleton is made is aragonite in most modern corals, but in the extinct orders the coralla may have been either partially or wholly calcitic.

Thus, the elements of a coral skeleton are: the epithecal wall, the radial septa and the horizontal floors (**tabulae**) which are secreted periodically as the cup enlarges and the polyp moves upward within it. These are present in all types of coral, but to varying degrees of prominence and in varying numbers. We shall see how this affects them as we consider each of the three orders separately below.

8.3 The colonial habit

We have already met, in the shape of the graptolites,

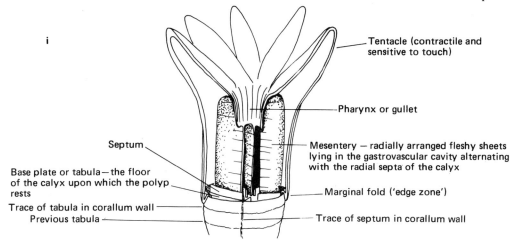

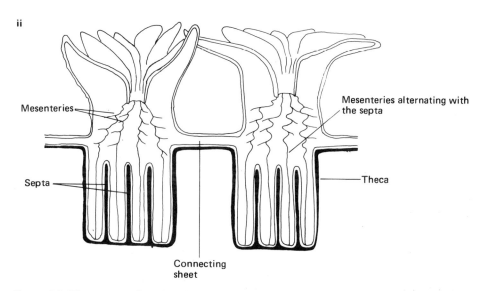

FIGURE 8.1. The anatomy of corals and its relation to hard parts. (i) a single calyx, (ii) coral polyps sharing a common coelenteron, as seen in many compound scleractinia. (Modified after Hyman.)

organisms which live as colonies. In these colonies, the organisms (**zooids** is a general term for them) are held together in a common skeleton. They are all descendants by asexual division of an original founder, and they may be physiologically linked to a greater or lesser extent. In some colonies the zooids may become specialized to fulfil particular functions. In these cases the individual appears to surrender its potential, independent viability, by dedicating itself to the 'greater purpose' of the whole: which, by virtue of its integration, functions as a kind of super-organism.

A degree of functional specialization is presumed for the autothecae and bithecae of dendroid graptolites (Ch. 4), and the control of thecal growth on side branches of cyrtograptids seems to demand a relatively high degree of physiological integration. Colonial integration is seen to varying extents in the compound corals of the three orders. This can be inferred from structural evidence in the corallum — such as, for example, a perforated epitheca. Closely packed polyps or even total reduction of the outer calyx wall all point similarly to increased coordination and interdependence. We shall see that such integration has reached a peak in modern corals, and may, in part, be responsible for their success.

But what are the advantages of the colonial habit? Increased skeletal size confers stability and also greatly improves the security of polyps against predators. Also, by colonial growth, the efficiency of feeding and respiration may be improved, either by concerted current induction, or merely by virtue of size and by being raised from the sea bed into stronger and cleaner water flows. In corals, however, specialization of polyp function is all but unknown.

8.4 Rugose corals

The Rugosa, so named for their wrinkled exteriors, are almost totally Palaeozoic and may either be solitary or compound (Fig. 8.2). In solitary species the

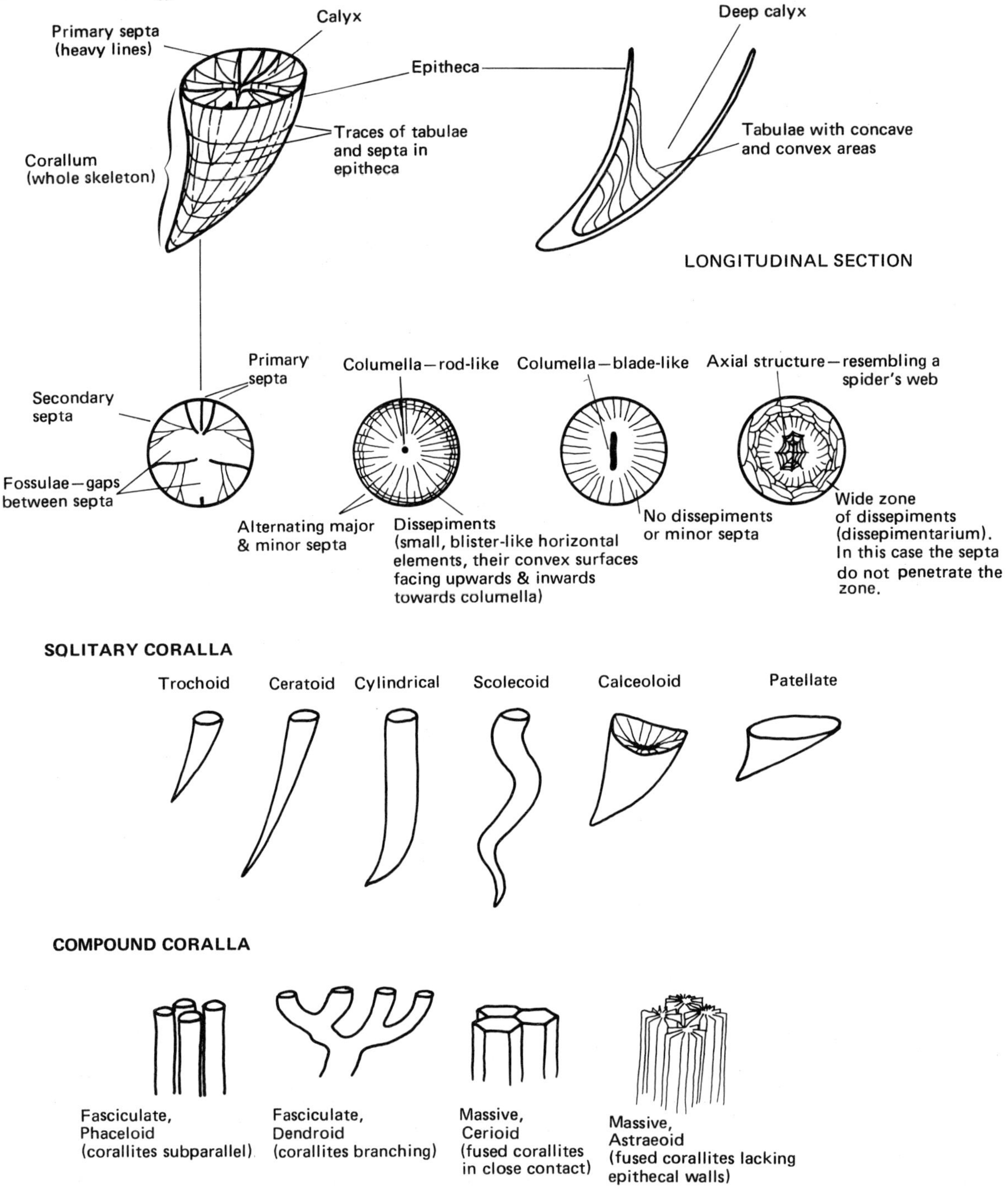

112   *Palaeontology — An Introduction*

FIGURE 8.2. Rugose coral morphology.

corallum is horn-shaped and the calyx divided by six main septa. These are inserted at four points along the periphery. As the coral grew it moved upwards, vacating the lower portions of the corallum, which were then closed off by tabulae. Such horizontal elements which occur in the peripheral region between septa, and do not extend across the entire corallum, are termed **dissepiments** (Fig. 8.2, 8.4). A wide zone of these is called a **dissepimentarium.**

When the original polyp budded to form a compound corallum, the individual calyces (now known as **corallites**) may either have fused together (forming a massive colony) or remained separate (to form a **fasciculate** colony). In the latter case, the separate corallites could either have lain parallel (termed **phaceloid**) or have assumed a more irregular pattern (termed **dendroid**). Examples of these morphologies are shown in Fig. 8.2.

Colonial integration was obviously higher in the massive type of skeleton. Packing of corallites was often so tight that they assumed polygonal outlines (termed **cerioid**). Perhaps the most highly integrated

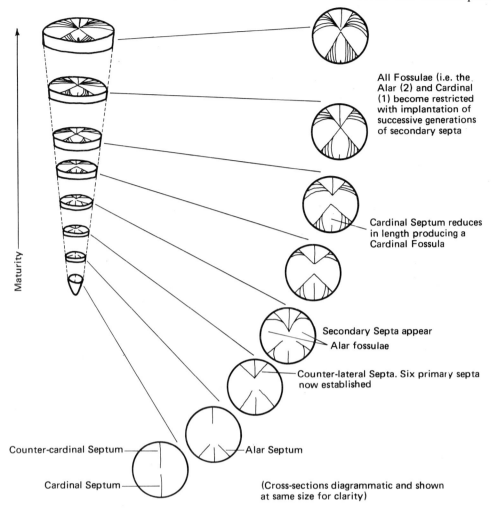

FIGURE 8.3. Septal ontogeny (developmental sequence) in a simple rugose coral (after Carruthers).

of all abandoned corallite walls altogether. Because of the star-like pattern made by the radiating septa, such corals are described as **astraeoid**.

As growth proceeded, metasepta were introduced between the prosepta in regular fashion. The process of insertion, which constituted the **ontogeny** of the individual coral (i.e. its development from inception to maturity), may be investigated by taking serial sections along the corallum (Figs. 8.1, 8.3). Here we see the six primary septa. Note how the metasepta are then inserted at four separate points, coming to lie on either side of the **cardinal septum** as well as on the 'counter' side of each **alar septum** (Fig. 8.3). Detailed analysis of similar sequences is vital to the fine taxonomy of these fossil corals.

The absence, reduction or parting of septa may give rise in some species to spaces which are known as **fossulae**. There may also be a space in the central area, though a central rod or columella may occupy this site. In the central area of some rugose corals a complex **axial structure** resembling a spider's web is seen. These are shown in Fig. 8.2.

8.5 Tabulate corals

These are a Palaeozoic group of exclusively colonial corals. They get their name because of the reduction of septa and the relative prominence of the horizontal elements in their structure. Although their colonies were often metres across, the individual corallites were small, measurable in millimetres. A degree of colonial integration was also probably attained, to judge from the closely packed corallites and the presence in some (e.g. *Favosites*, Fig. 8.6), of **mural pores** which serve to interconnect them. Even in fasciculate forms such as *Syringopora* (Fig. 8.6), cross-linking tubes were present.

Tabulate corals often took part in the formation of Palaeozoic reefs. In this they were accompanied by the stromatoporoids (see below) to which they were subordinate in those reefs which grew in vigorous, shallow water. Deeper-water reefs tended to have a greater proportion of tabulates (Fig. 8.11), and forms such as *Halysites* (Fig. 8.6) were characteristic of areas with very high sedimentation. Their mode of

114 *Palaeontology — An Introduction*

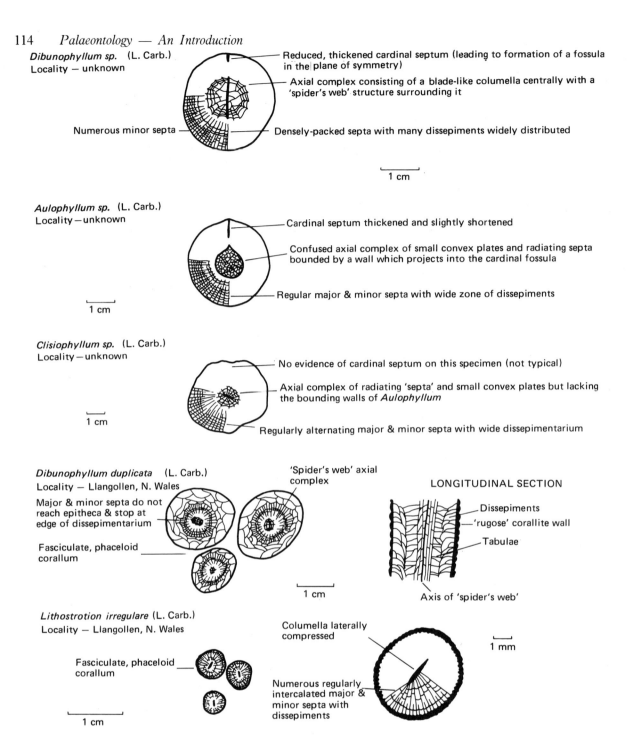

FIGURE 8.4. Solitary and compound rugose corals of the Carboniferous Limestone. (All from original specimens.)

construction allowed fast growth to keep ahead of burial. Moreover, corallites at the centre of the colony could clean themselves and dump the material into the shafts between the interlocking **palisades**.

Tabulates were unable, however, to cement themselves to their substrata — a disability which put them at a severe disadvantage in reef habitats.

8.6 Scleractinian corals

The spectacular geological history of the Scleractinia (Fig. 8.7) stretches through Mesozoic and Caenozoic to the present, when they are possibly more successful than ever. Like the Rugosa (from which they may have evolved), they may be either solitary or colonial and they also have six prosepta in the calyx. Subsequent metaseptal insertions, however, are in multiples of six.

Solitary polyps may be large (up to 25 mm across), but in compound forms they average about 1–3 mm in diameter. The scleractinian polyp overhangs its calyx, and this overlapping fold of body wall allows the corallum to conform closely to the substrate and become cemented to it.

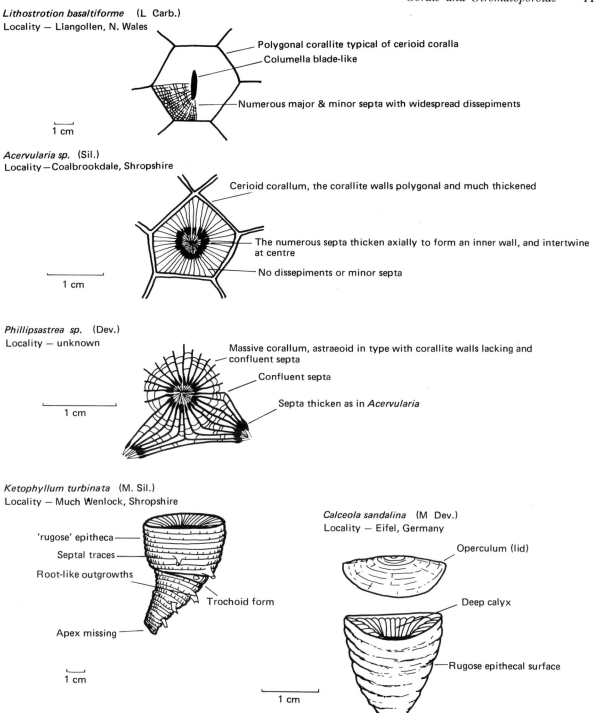

FIGURE 8.5. Rugose corals of the Upper and Lower Palaeozoic. (All from original specimens.)

The coralla of scleractinian corals are light and delicate. Solitary types tend to be horn-shaped, and compound ones show a high degree of zooid integration. In the brain coral *Diploria*, for instance (Fig. 8.7), the polyps are fused in meandering chains, their adjacent mouths lying in a valley between common tentacles and leading to a shared coelenteron (Fig. 8.1).

So the scleractinians owe their success partly to being able to cement themselves down, and partly to their ability to fuse polyps. But the major element in their success has been their symbiotic relationship with non-motile dinoflagellates (see Ch. 10). These are photosynthesizing micro-organisms which live in the endodermal cells and which are usually referred to as **zooxanthellae.**

Being plant-like, these manufacture food from CO_2 and sunlight, and they probably donate some of the

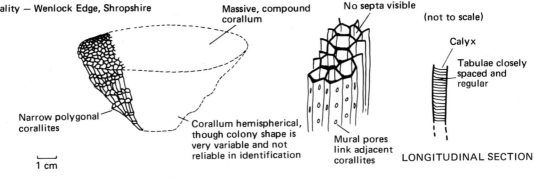

FIGURE 8.6. Some common tabulate corals. (All from original specimens.)

products to their hosts. But the dinoflagellate and the coral both require the element phosphorus in order to grow, and this element is rare in sea water. It appears that the two partners recycle their phosphorus from one to the other, hence ensuring that there is little lost to the environment. Most significant of all (from a geological viewpoint) is that the removal of CO_2 by photosynthesis eases the precipitation of calcium carbonate. Scleractinia can actually calcify ten times more quickly by day than by night, when photosynthesis cannot occur.

Of course, this symbiosis can only work if the coral lives in the photic zone of the upper ocean. In fact, it is found that corals which grow in reefs (known as **hermatypic** corals) are also those in which the symbionts are found. Hermatypic corals are ecologically restricted to waters of precise salinity, temperature and clarity in tropical seas (Fig. 8.12). **Ahermatypic** corals (which are non-reef-forming and lack zooxanthellae) are much less particular. They can exist at depths of 6000 m and survive temperatures as low as 1°C. They are mostly solitary, and some even live in the icy but clean waters off the north-west of Scotland.

8.7 Geological history and stratigraphic value

The earliest corals, the tabulates, may possibly originate in the Cambrian: but even their first

Corals and Stromatoporoids

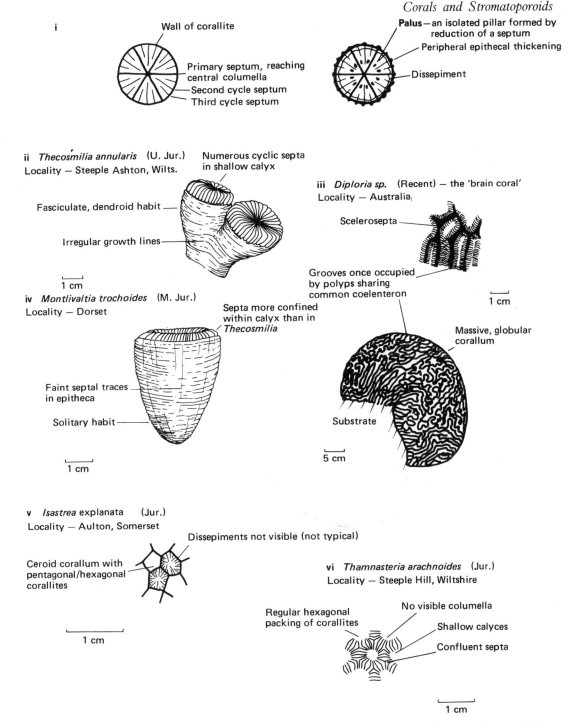

FIGURE 8.7. Scleractinian corals. (i) General morphology (after Moore, Lalicker and Fischer). (ii)–(vi) typical examples (after original specimens.)

unequivocal appearance in the Ordovician slightly predates that of the Rugosa. They became important in the Silurian and Devonian, when they were prominent in reef habitats. After this they declined and were extinguished at the end of the Permian.

Most tabulates are of no special stratigraphic value, though certain forms are useful markers, e.g. *Pleurodictyum* is restricted to the Lower Devonian.

Rugose corals, originating in the Ordovician, also rose to prominence during the Silurian. They continued to diversify through the Devonian (when the slipper coral *Calceola* (Fig. 8.5) appeared, a good marker for the Middle Devonian). In the Carboniferous they reached an acme of diversity, and are used in the Carboniferous Limestone as zonal indices (Fig. 8.9). However they, like the tabulates, eventually died out — not, in this case, at the end of the Permian, but shortly after.

The Scleractinia, deriving possibly from the Rugosa, dominated the Mesozoic and Caenozoic

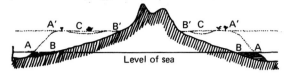
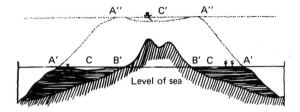

(From Darwin's original diagrams)

FIGURE 8.8. Charles Darwin's explanation for the origin of the coral atolls of the Pacific. In the top diagram, the volcanic island has a fringing pavement of corals growing around its flanks. As the sea level rises relative to the island, coral growth continues to keep pace with it forming first a fringing barrier reef (A'–A') with a lagoon separating it from the half-submerged island. Continued relative rise of sea-level eventually submerges the island: but continuance of coral-growth ensures that the ring of coral which had surrounded the old volcano still maintains its position at sea level (A"–A") forming an atoll.

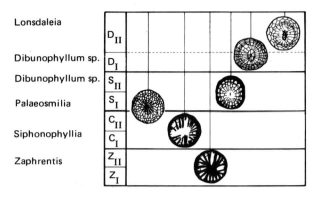

FIGURE 8.9. The use of rugose corals in the zonation of the British Carboniferous Limestone. The zone code-letters are those erected by Vaughan in 1905. Some are based on the coral's generic name (e.g., 'Z'), but others are based on outmoded generic names or upon brachiopods (e.g. 'S', for *Seminula*, now itself re-named!).

scene. They are not, unfortunately, of much stratigraphic use, but they can build reefs ranging from very small **atolls** (Fig. 8.8) to massive structures on the scale of continents. The Great Barrier Reef, off the east coast of Australia, is visible from space!

Within these diverse and often enormous structures, the plasticity of coral growth and the distinct ecological zones into which they fall are striking. Recognition of such ecological zonation in fossil reefs is one of the most exciting and challenging parts of modern palaeontology, in an area where this subject grades imperceptibly into the allied discipline of sedimentology. An understanding of coral ecology is important in the interpretation of their environment, and the great reservoir potential of reefal build-ups has led the oil industry to finance much research into this complex field.

II. STROMATOPOROIDS
Phylum Stromatoporoidea (Camb.–Cret.)

Although stromatoporoids are probably more closely related to sponges than corals, they are described here because they tend to be found alongside tabulate or rugose corals in Palaeozoic reefs. The biological affinities of this perplexing, extinct group have only recently been established, yet even now there is no unanimous agreement. Maybe when there is, the Phylum Stromatoporoidea, as it now stands, will be scrapped, and the fossil forms grouped in the Class Sclerospongia of the Phylum Porifera (the sponges). We shall have to wait and see.

Stromatoporoids range in size from thick laminae metres across, and domical skeletons a metre tall, to the tiniest of minor encrustations. Their external form is a poor guide to their taxonomy, since it was apparently subject to environmental, rather than genetic control. To enable adequate identification, the skeletons must be examined using a microscope and thin sections cut both vertically and horizontally (Fig. 8.10).

A vertical section reveals that the skeleton (called the **coenosteum**) was coarsely laminated. These broad layers are known as **latilaminae** and record growth periods, like the rings of a tree. But each of these is constructed of very fine horizontal and vertical components.

In some, the horizontal elements are like tabulae, solid calcareous floors, separated by pillars to form **galleries**. The whole structure resembles a microscopic multi-storey car-park. In others, the horizontal components are also bars, so that the coenosteum looks more like a mass of scaffolding.

On the latilaminar surface, small hummocks called **mamelons** may be seen, and radiating from their summits one sometimes observes radiating grooves called **astorhizae**. The function of these, and indeed the whole nature of the soft parts of stromatoporoids, is a matter for conjecture.

Mesozoic stromatoporoids were not generally reef-builders, but Palaeozoic representatives were often their dominant components. Difficulties in identification, however, have rendered their practical stratigraphic use next to impossible.

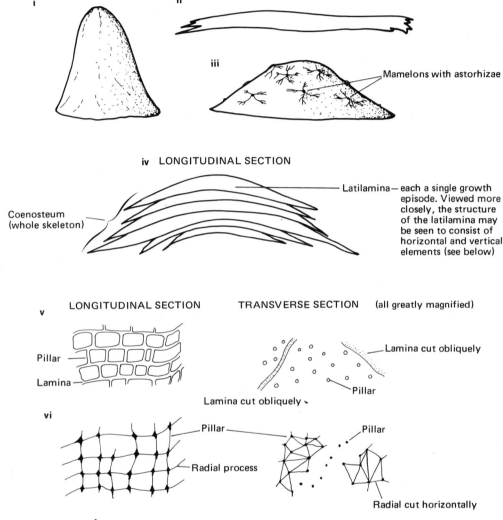

FIGURE 8.10. Morphology of stromatoporoids. (i) High-domical coenosteum; (ii) laminar coenosteum; (iii) low-domical coenosteum. Shape of coenostea bears only the poorest of relationships with true taxonomy and so is not a useful feature in identification. Note that (iii) bears small pimple-like eruptions called 'mamelons'. These have a central pit from which channels radiate over the coenosteal surface. These are the 'astorhizae'. Mamelons and astorhizae do not occur in all stromatoporoid species. (iv) Vertical section of a low-domical stromatoporoid showing the latilaminae. Each represents a period of growth, like a tree-ring: though its chronological significance has yet to be established. The microstructure of the latilamina is shown in (v) and (vi). In (v) the horizontal elements are floors and they are separated by pillars like a miniature multi-storey car-park. In (vi) the horizontal and vertical elements are bars, forming the so-called 'hexactinellid' network, rather like scaffolding, with all diagonal elements in the horizontal plane. (Modified from Mori, 1969.)

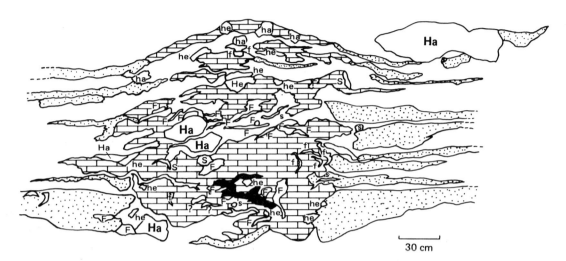

FIGURE 8.11. Transverse cliff-section of a Llandovery (L.Sil.) reef composed of tabulate corals and stromatoporoids. Ha = *Halysites*; He = *Heliolites*; F = *Favosites*; S. = stromatoporoid. Dotted shading indicates bioclastic limestone. Brick shading denotes 'reef rock', a complex limestone made up of calcareous algae, micrite and some crinoidal material. Black/unshaded areas signify fine green mud. Locality — S. of Kneippbyn, Gotland, Sweden. From Nield, 1981.

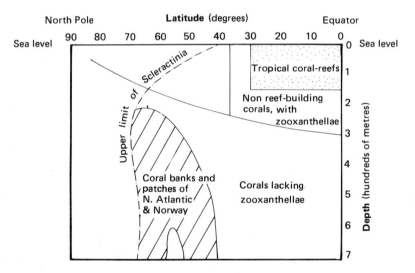

FIGURE 8.12. Depth and latitude-ranges of modern corals and coral-reefs. (Modified from Teichert, 1958.)

9.
Vertebrates

9.1 Introduction

Animals with a dorsal nerve cord and an axial stiffening rod are classed as chordates (Phylum Chordata). 'Primitive' chordates have only a flexible bar which lies ventrally to the nerve cord without enclosing it. This bar is called a **notochord**, and by becoming bony or cartilaginous, as well as by enlarging so as to envelop and protect the nerve cord, it evolved into the jointed **spinal** or **vertebral column** seen in the vertebrates (Subphylum Vertebrata — Table 9.1). Only they, of the Phylum Chordata, have a fossil record, and an exceedingly complex one it is.

Firstly, the internal skeletons of vertebrates may be cartilaginous rather than bony, and thus are not likely to be preserved. Also, disarticulation of the bones after death is very common, making complete finds very rare. Nevertheless, the excitement of finding the meanest vertebrate relic (a shark's tooth, some fish-scales or a reptilian vertebra) is easily matched by the strong fascination of an evolutionary story which, ultimately, leads directly to ourselves.

To do justice to the Subphylum Vertebrata would require a chapter as long as the rest of this book; nature does not make her units in neat, comparable packages. So, to try and make sense of the story, we shall concentrate on the vertebrates' long march from the water's edge to a full and independent terrestrial life.

9.2 The vertebrate plan

Vertebrate skeletons are internal, bilaterally symmetrical and made of either cartilage or bone — though only the latter provide good fossils — and the skeletal elements articulate at joints which do not hold together after death. The skeleton may be thought of in terms of an **axial** portion (spine, bearing the skull at the front end), two pairs of **limbs**, placed one at each end of the spine, and two plate-like **girdles** (Fig. 9.1). These elements are always present: in fish, frogs, dinosaurs, whales, bats, humans and gerbils, changing only in their proportions and in certain details.

During vertebrate evolution, the development of the limbs and the addition of a jaw to the skull are the most significant changes in the early plan. But, as we shall see, the conquest of land also necessitated radical alterations in soft-part anatomy, physiology and reproductive habit which are not necessarily clear from the fossil evidence alone. For such information we must turn to living representatives of these ancient vertebrate classes.

TABLE 9.1. *A Classification of the Vertebrates*

Phylum	Subphylum	Superclass	Class	
CHORDATA	VERTEBRATA	Gnathostomata (Jawed vertebrates)	Placodermi	Aberrant Palaeozoic jawed vertebrates
			Chondrichthyes	Cartilagenous fishes (eg. sharks, dogfish)
			Osteichthyes	Bony fishes (eg. salmon, trout, cod etc)
			Sarcopterygii	Lobe-fin fishes and lungfish
			Amphibia	Tetrapods lacking an amniotic egg
			Reptilia	Tetrapods with amniotic egg, but no fur or feathers
			Aves	Amniotic egg, feathers — ie., the birds
			Mammalia	Hair, nursing habit and advanced brain
		Agnatha (Jawless vertebrates)		

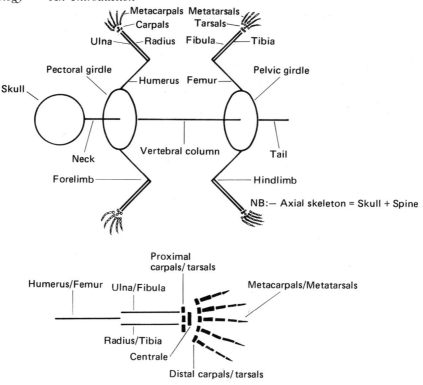

FIGURE 9.1. The vertebrate body-plan and the five-rayed or pentadactyl limb.

9.3 Jawless fishes

We are fortunate that some (albeit rather untypical) jawless vertebrates have survived in such creatures as the lamprey, *Lampetra* (Fig. 9.2). It is an eel-like creature, lacking paired fins but possessing a functional notochord. The skeleton is cartilaginous. Most noticeable, apart from its jawless, sucker-like mouth, are the **gill-slits** which run down the sides of the anterior end of the body. Each of these allows water, entering the mouth, to pass out over the blood-gorged gills and so exchange bodily carbon dioxide for oxygen. Each of these gill-slits is supported by skeletal **arches** — these will become important later in our discussion. Also of some future significance is the **nasal opening** on top of the head. In the lamprey it leads to a sensory sac only, but in more 'advanced' forms it will be pressed into service for respiration in air.

The most notable fossil agnathans are the **ostracoderms**, armoured fishes from the Palaeozoic. Their earliest remains date from the Ordovician (of Colorado), where scales and teeth have been found. But it is only in the U. Silurian that recognizable body fossils first turn up.

These are beasts like *Cephalaspis* and *Psammolepis* (Fig. 9.2 i, ii). They, like the lamprey, had no jaws and no true paired fins, and many had the single nostril on top of the head. But their heavy **head shields** point to predation and the need for protection from some early enemies. These could well have been giant arthropods, like the eurypterids.

Geological and biological evidence strongly suggests that these early vertebrate creatures lived in fresh water. But, once established, they spread rapidly into the seas, so that their next stages of development are more widely recorded in the marine sediments of the Devonian — a period which has become known as 'the age of fishes'.

9.4 The origin of jaws and lateral fins

Lack of jaws restricted early fishes to a life on the sea floor. To assume predatory roles, these creatures had to take on torpedo-like shape for greater swiftness, develop jaws for grasping prey and lateral fins to prevent rolling movement now that their old, flattened form had been abandoned.

In sharks and dogfish (Fig. 9.3), jawed forms with cartilaginous skeletons, forward propulsion is provided by the lashing of the tail. But since their bodies are denser than water, side fins, in addition to stabilizing these fish, must also act as 'aerofoils' to keep them off the bottom. Sharks swim, then, like an aeroplane flies; bony fishes, on the other hand, swim like powered airships, being naturally buoyant. Incidentally, sharks also have to keep in motion because it is only their own velocity which maintains the flow of water over the gills.

To form jaws, the anterior gill-arches were pressed into service. And as they became modified, so the number of gill-slits was reduced. The slit nearest the head was lost when its arch formed part of the base of

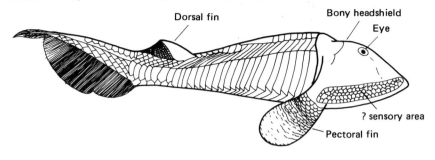

i A Devonian agnathan, the heavily armoured *Cephalaspis* (after Stensiö)

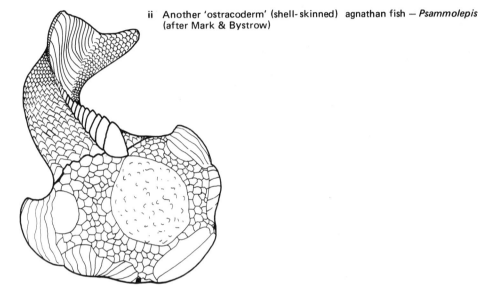

ii Another 'ostracoderm' (shell-skinned) agnathan fish — *Psammolepis* (after Mark & Bystrow)

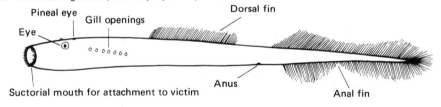

iii A modern agnathan, the lamprey (*Lampetra*)

FIGURE 9.2. Agnathan or jawless vertebrates. (Not to scale.)

the cranium. The second vanished when the arches supporting it rotated to become the upper and lower jawbones (Fig. 9.3 ii). The third slit disappeared when its arch moved closer to the jaw in order to support it.

The origin of the jaw by gill arch modification is evident in the skull of the Devonian shark *Cladoselache*. In its jaws and lateral fins, this fish was constructed on what became the standard vertebrate plan. But this blueprint was not established straight away. Strange, 'experimental' fishes (grouped together as **placoderms**) are also found, which differ quite strikingly. Examples of this group include the **arthrodires** ('jointed necks') — fish which could grow to a terrifying 10m and had massive head-shields (Fig. 9.4).

9.5 The origin of bone and lungs

There is a danger in broad treatments like this to give the (quite false) impression that the changes which occurred took place progressively and purposefully, with something particular in mind, or as though guided by some spectral hand. For example, it is wrong to assume that the first vertebrates were cartilaginous and were gradually improved by the invention of bone. The origin of bone in vertebrates is an event which took place very early in their evolution. *Cephalaspis*, for example, had a bony head shield, yet it is, from an evolutionary point of view, a more 'primitive' form than the cartilaginous shark. So the origin of the bony fishes (**Osteichthyes**) was really the

124 *Palaeontology — An Introduction*

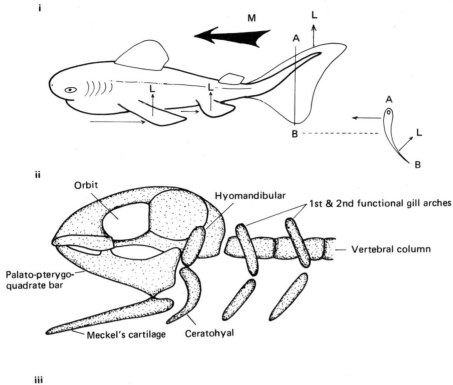

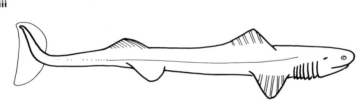

FIGURE 9.3. Locomotion and jaw suspension in cartilaginous fish and an early example of the Chondrichthyes. (i) As the dogfish moves ahead (large arrow, M) the pectoral and pelvic fins act as 'wings' to generate lift (L) from the motion of the water against them. Side-to-side lashing of the tail (section A–B) also creates lift. As the tail moves to the left, lift is generated in its wake. On the return stroke, the fin bends the other way and so generates lift in both directions. Lift is vital to cartilaginous fishes of this kind since they do not have buoyancy and sink when at rest. (ii) The jaw of a dogfish is formed between the palato-pterygo-quadrate bar and Meckel's cartilage. These are the modified remnants of the first gill arch seen in jawless fish. Supporting them is another modified gill arch, formed of the hyomandibular and ceratohyal. Unaltered gill arches, still functional in supporting the gills, follow on. (iii) A Devonian shark (after Dean). It had broad, paired fins and its jaws were very obviously homologous with the gill bars.

re-introduction of bone into the vertebrate scheme. Moreover, it took place not by the ossification of pre-existing cartilage, but the development of bony veneers upon them.

The introduction of bone into the main skeleton was, in a sense, an important 'pre-adaptation' to life on land, where the skeleton would have to support much more body weight than in water. This expression 'pre-adaptation' must not lead you to conclude that these creatures 'knew what they were doing' when they became bony. 'Pre-adaptation' simply implies that a new feature, evolved for one immediately pressing reason, was, later in geological time, of advantage towards some other purpose. Foresight never enters into evolution.

Another major step forward which was taken in the Devonian was the development of **lungs** for breathing air. These structures originated as sac-like pouches on the underside of the throat. They were paired, and they developed so as to allow fish to live in temporary bodies of water prone to seasonal drying. By possessing lungs, these fish could live buried beneath the mud during dry seasons, breathing air in a state of suspended animation, while awaiting the return of the water.

Three species of these **lungfish** still survive, such as the African genus *Protopterus* and the American genus *Lepidosiren* (Fig. 9.4 ii). In these forms, the nostril (see above) has at last become part of the respiratory system, so that breathing can take place without the need to open the mouth.

Lungs, therefore, actually developed first in aquatic

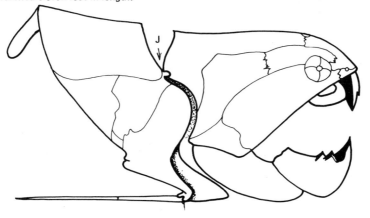

i Side-view of the headshield of the arthrodire *Dinichthys*. This giant inhabited Devonian seas and may have reached 30 feet in length. The name 'arthrodire' refers to the jointed neck (J) which allowed limited articulation of the head relative to the body. The armoured portion shown may have reached eleven feet in length.

ii Side-view of the S American Lungfish *Lepidosiren*. The lungfish are lobe-fin or sarcopterygian fishes (see below) but they have adopted an eel-like form and have reduced their fins considerably.

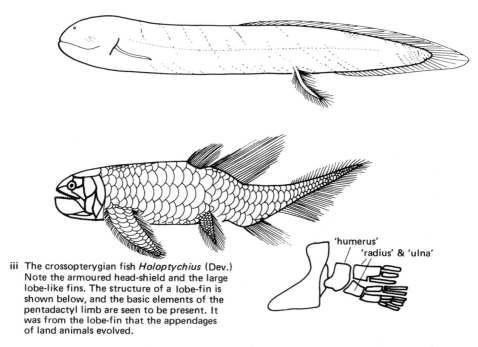

iii The crossopterygian fish *Holoptychius* (Dev.) Note the armoured head-shield and the large lobe-like fins. The structure of a lobe-fin is shown below, and the basic elements of the pentadactyl limb are seen to be present. It was from the lobe-fin that the appendages of land animals evolved.

FIGURE 9.4.

animals. But it was an important development for all fishes, because these paired bladders are used today in wholly aquatic fishes as **swim bladders**. These are hydrostatic organs which, by flooding or emptying, adjust the animal's buoyancy and allow it to rise or fall in the water.

9.6 Amphibians and terrestrial vertebrates

With bony skeletons, paired fins and primitive lungs all developed in the Class Sarcopterygii (Table 9.1), the elements were assembled which would be needed for life on land. One group in this class includes the lungfish; another consists of the **lobefins**, fish which may have been the direct precursors of four-footed land animals.

These fishes, plentiful in the Devonian (Fig. 9.4 iii) and still with us today in the form of the recently discovered *Latimeria*, possessed functional respiratory nostrils, lungs and paired fins, whose construction bears a directly analogous relationship with the limbs of tetrapods.

Early amphibians, such as those from the Devonian of Greenland (Fig. 9.5), bear obvious fish-like characteristics. But although the further development of the lobe fin into the first limb enabled locomotion on land,

126   *Palaeontology — An Introduction*

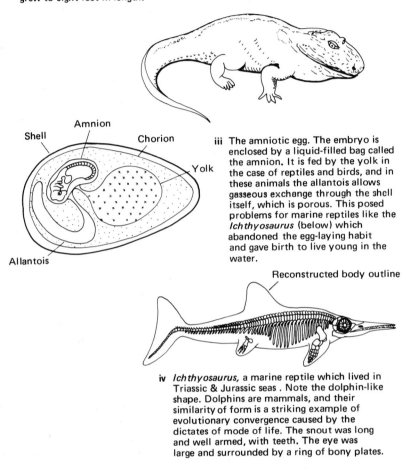

FIGURE 9.5. Early amphibia, an amniotic egg and a marine reptile.

their lungs (like those of modern amphibians) were inefficient and needed to be supplemented by a glandular, respiratory skin.

So their skin needed to be kept moist to work properly as an organ of gaseous exchange. Another factor tying the amphibia to water was their reproductive system. Like that of frogs today, this involved the external fertilization in water of gelatinous eggs, and the need for a larval tadpole stage.

But, despite these difficulties, the Carboniferous saw the development of some very large amphibia, notably *Eryops* (Fig. 9.5 ii). This 2-m-long creature probably spent very little of its life out of the water, however.

With the coming of reptiles, we see the elimination of many amphibian shortcomings as land animals. Firstly, reptilian lungs are efficient enough to function without the need of respiratory skin as a 'back-up'. Secondly, fertilization is internal, involving copulation between male and female and the production of a sealed egg. This egg can breathe air through its porous shell, and is endowed with enough food (yolk) to fuel development through the completion of embryonic growth (Fig. 9.5 iii).

As with lungs, however, it appears that the laying of sealed eggs on land was first developed by aquatic mammals. Certain reptiles today (the turtle and the crocodile, for example) come ashore to lay their eggs. It is a strategy which is of advantage to them because the eggs are less vulnerable and the need for a tadpole stage is eliminated. Such a life cycle is not affected if the waters should dry up — which would, of course, be a complete disaster for tadpoles.

But the sealed egg laid the way open for the

colonization of land. This process produced spectacular results in the dinosaurs, a superbly successful group of extinct reptiles which rose to dominate the habitat throughout the Jurassic and Cretaceous periods. They underwent a huge diversification into carnivores, herbivores, flying forms — even apparently reversing the landward trend and becoming fully aquatic.

Marine reptiles, such as the dolphin-like *Ichthyosaurus* (Fig. 9.5 iv), were, interestingly, posed with a severe problem in the sealed egg. For these eggs need to breathe air, and if laid in water will actually 'drown'. The ichthyosaur therefore abandoned the egg and gave birth instead to live young. We know this, since pregnant females of the species have been fossilized with their unborn young inside them. Giving birth to live young in this way is called **vivipary**.

There is not the space to enter here into the many engaging problems which surround the dinosaurs. There is a strong possibility, for example, that they were warm-blooded. And what of their extinction? Such topics continue to sell paperbacks and newspapers like no other part of the science. But with regard to their extinction, these mostly miss the really important point, which is, why did similar extinctions occur in so many other groups of animals as well? The problem of mass extinctions is treated further in Chapter 12.

But another intriguing theory has recently hit the headlines, one which lies closer to our purpose in this chapter: namely, that dinosaurs gave rise directly to the birds. A close affinity between reptiles and birds has always been accepted, but recent discoveries have made it possible that the birds actually derived from a truly 'dinosaurian' stock, rather than from a common ancestor of both groups.

The earliest birds may well have been able to catch worms, but they were certainly no good at flying. The famous Jurassic bird *Archaeopteryx* (Fig. 9.6 i) possessed feathers, but they were probably developed for insulation rather than any aeronautical purpose, since the musculature and bone-structure of this creature were quite inadequate for it to have performed powered flight. Instead, the feathery wings of *Archaeopteryx* may have enabled gliding or pouncing, and may also have been used in the entrapment of prey.

But *Archaeopteryx*, while undoubtedly 'avian', has many reptilian features. These include teeth, a long tail, clawed digits on the forearms; its bones are also solid. The bones of modern birds, on the other hand, are hollow — to save weight — and are strengthened by an internal strutting system (Fig. 9.6 iii).

Once again we see that the process of evolution does not assemble all the characteristics of any major group all at once. Each feature is developed for its own reasons, reasons often removed from its eventual function.

9.7 Goodbye to the egg — the origin of mammals

Mammals are distinguished (and get their name) from the **nursing habit**, i.e. the suckling of the young on maternal milk produced by **mammary glands**. Like birds, they maintain a high blood temperature and can therefore be active in (more or less) any climate. But they, uniquely, possess hair (fur) and a complex brain requiring a relatively larger cranium in which to house it. These modifications produce vigorous, intelligent animals, efficient in pursuit and skilled in manipulation.

We can trace the mammals to a group of mammal-like reptiles called **theraspids**, which became widespread in the Permo-Trias, well before the great age of the dinosaurs. Their skeletal similarity to mammals is seen in their possession of teeth which are differentiated into different types (molars, incisors, etc.); the detailed make-up of their jawbones; the nature of the joint between skull and spine; and the changed attitude of the legs relative to the body. These characteristics are either fully 'mammalian' or else they show a tendency towards the mammalian condition.

These creatures, from the Karroo Beds of South Africa may (who knows?), have possessed hair, and even nursed their young. These sorts of things one cannot tell from bones alone, and that, unfortunately, is all we have. The theraspids did not survive the 'age of reptiles'. Presumably they did give rise to mammalian descendants, but the sad fact is that, in the Jurassic and Cretaceous, the mammals went into a 'preservational eclipse'. Indeed, such is the scarcity of Mesozoic mammal remains that the great American vertebrate palaeontologist A. S. Romer was moved to suppose that they could all be fitted quite comfortably into 'an old-fashioned Derby hat'.

But from these fossils we can be sure that early mammals grew no larger than the rabbit. Their scanty remains have turned up in Britain in the Stonesfield Slate (M. Jurassic), the 'dirt-beds' of the Purbeck (U. Jurassic) and in the Wealden Series (L. Cretaceous). Earliest of all, and the most enigmatic, are some remains reported from the Rhaetic Series of the Upper Triassic. In a solution pipe below the basal unconformity, where the Rhaetic rests upon the Carboniferous Limestone in the Bristol and South Wales region, fragments of '*Hypsiprimnopsis rhaeticus*' were discovered. Unfortunately, the specimens are now lost.

Although mammal-like reptiles are extinct, reptile-like mammals are not. The **monotremes** (Table 9.1), exemplified by the duck-billed platypus *Ornithorhynchus*, have hair and suckle their young, but not before giving birth to them in eggs! All other mammals are viviparous, however, and as evolution progressed the tendency was for the foetus to spend ever longer within the security of the mother's body.

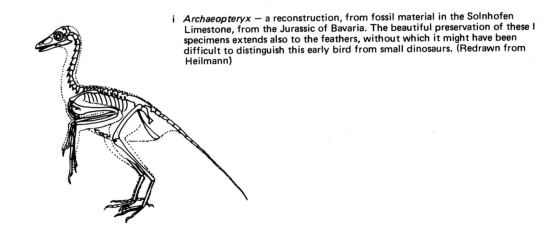

i *Archaeopteryx* — a reconstruction, from fossil material in the Solnhofen Limestone, from the Jurassic of Bavaria. The beautiful preservation of these specimens extends also to the feathers, without which it might have been difficult to distinguish this early bird from small dinosaurs. (Redrawn from Heilmann)

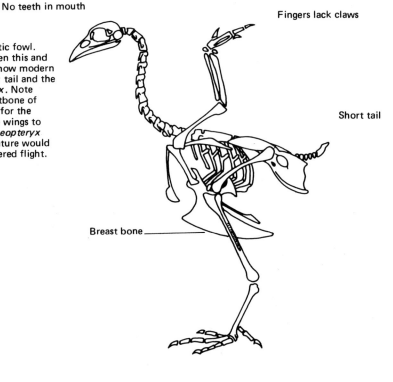

ii Skeleton of *Gallus*, the domestic fowl. Look at the differences between this and its Jurassic predecessor. Note how modern birds lack teeth, the long bony tail and the clawed digits of *Archaeopteryx*. Note also the immense keeled breastbone of the modern bird. This is there for the huge muscles which power the wings to attach to. Its absence in *Archaeopteryx* strongly suggests that this creature would not have been capable of powered flight.

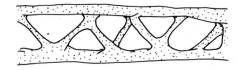

iii The internal structure of bone from the wings of a bird. This cross-strut arrangement is extensively used by aero-engineers, and is known as the 'Warren's Truss', a design which has been found to maximize strength and at the same time minimize on materials (and, therefore, weight)

FIGURE 9.6. The earliest bird, a modern example and some typical bird bone.

In other words, pregnancy (the gestation period) was lengthened.

In marsupial mammals, the mother gives birth to a very poorly developed, foetal creature which must grope its way to the pouch (the **marsupium**) and the teat it contains, in order to develop further. Many fossil marsupials are known, notably from the U. Cretaceous of the USA and the Tertiary of S. America.

The remaining mammals, the so-called placental type, show the next and final advance in the chain.

These forms (which include mammals from Aardvarks to Zebras, taking in elephants, rodents, carnivores, bats, whales, monkeys and seals) have a very prolonged gestation, made possible by the **placenta**. This is formed from one of the old egg membranes, the **allantois** (Fig. 9.5 iii), which has become attached to the wall of the uterus and permits the blood systems of mother and child to exchange oxygen, carbon dioxide and food without actually mixing. (Mixing, of course, would be disastrous. The mother's immune system would reject the child as foreign tissue, just as transplant patients will reject their donated organs without special drugs to suppress their natural immunity. Moreover, the mother's blood-group and that of her child may not be the same.)

The earliest placentals were small, insectivorous beasts, like shrews and hedgehogs, and they first appeared in the late Cretaceous. Mammalian radiation took off — literally, in the case of the bats — in the Tertiary. And despite the huge range of life forms which resulted in response to the new modes of life which were adopted, the basic five-fingered or **pentadactyl limb** is present in each (Fig. 12.1). The degree of elongation or suppression in each of its component parts is all that masks their comparative homology, from bat's wing to dog's paw, whale's fin and horse's hoof. And the pattern and range of mammalian diversity, despite its incredible width, was fully established (in something very like its modern scope) by the beginning of the Eocene.

9.8 The process of man's emergence

The illumination of man's place in nature is usually billed as a major motive in palaeontology (take the first chapter of this book, for instance). And although our broad position and the processes which encouraged our development are fairly well understood, the precise details of our evolution over the last 3 million years or so are very sketchy. Indeed, they seem to change after every field season in the East African Rift Valley. For it is here (or, more precisely, in the Olduvai Gorge) that river sediments and volcaniclastic deposits have preserved this latest period of hominid evolution.

Generally, a denizen of forest and plain rarely becomes fossilized, and this, together with the ramifying complexities of 'main line' and 'offshoot' in the early history of our kind, makes the few dingy details at our disposal look very crude and controversial. But what is it that makes a human, and what processes guided the changes and set them in motion?

Primates (monkeys, apes and man) divided quite early in their history into two basic types. It was from the African/Asian or 'Old World' stock that the so-called great apes developed. These **anthropoid** or 'man-like' apes are very similar to man, both in skeletal structure and in the amino-acid sequences found in their enzymes and proteins. Indeed, the differences in flesh and bone between great apes and man are of relative proportion only. This is what unites us with the apes, not what sets us apart, however. And we still have not answered the question 'why did we change?'

The human body is not functionally specialized, but many of its characteristics were developed in our arboreal ancestors. A hand with an opposable thumb for grasping is necessary for quick, acrobatic movement through the canopy. And such skill also requires good coordination of muscles, and three-dimensional vision.

Man's direct ancestors came down from the trees as the forests were gradually replaced by grassland in the Tertiary. It became necessary to see across large distances, necessitating an upright posture. The shambling, knuckle-dragging gait of the ape was exchanged for an erect, bipedal stance, demanding major skeletal alterations.

The pelvis, for example, needed to widen and become shorter to support the weight of the body. Also, the spine had to enter the skull from below rather than behind, so the hole known as the **foramen magnum**, through which the nerve cord enters the brain, migrated underneath the cranium. In some ways our bodies are not yet properly adapted to this posture. Think of this when you suffer from varicose veins and lumbago!

In 1976 the fossil *Ramapithecus* was discovered in Pakistan. He may have been more ape than man, but the feeling is that he represents the earliest known stage in the change from tree- to plain-dwelling man, from stooping to erect stance. Between 8 and 6 million years ago the hominids as a group became distinct from the great apes. The skeleton came to look more 'human' and cranial sizes increased dramatically. The fossil evidence suggests that erect posture considerably predated brain enlargement, and biology tells us that it was much the harder to achieve. Our large brains (and our naked bodies) are preserved into adulthood by neoteny (subsection 6.22.2), the brain being expanded by prolonging high foetal growth rates and incurring no major structural changes. But the pelvic and pedal modifications which enabled us to stand up straight involved drastic redesigning. Perhaps we should be less proud of being clever and rejoice a little more in the ability to stand.

The major features of our species were fixed in the U. Pleistocene, and all our subsequent development has been cultural, not biological. The men and women of the stone ages were basically not the slightest bit different from modern examples of the species, so we should not be surprised when archaeology suggests that these people were ingenious, highly skilled, artistic and knowledgeable in astronomy and geometry. For we are no longer considering yet another mysterious life-form of the ancient past. We are looking in the mirror.

10.
Microfossils

10.1 When is a fossil a microfossil?

Although the term 'microfossil' might seem self-explanatory, things are not as simple as they appear. Firstly, it is not a biological term, and animal and plant remains of diverse origins can be grouped together as microfossils. If we were to take the groups described in the other chapters of this book and lump them together as 'macrofossils', we would be doing them as much justice as the term 'microfossil' does to those groups outlined below.

So the grouping is biologically abitrary. Furthermore, although some microfossils are the remains of very small organisms, many are merely the disintegration products of large ones. Sponges may fall apart into their component **spicules**, for example. Worms may decay completely away, leaving only their jaw apparatuses or **scolecodonts** to be discovered. Also, the **spores** and **pollen** which constitute such an important part of micropalaeontology are merely microscopic germinal cells produced by large plants.

The final irony in this discrimination based on size is that some representatives of microfossil groups are very large indeed. **Foraminifera** may often reach several centimetres in diameter!

The fossil record contains many fossils whose biological origin is a complete mystery. This is nowhere more true than in micropalaeontology. The **conodonts**, which were probably the component parts of some extinct, otherwise soft-bodied animal, have puzzled scientists since their discovery over a hundred years ago. But this puzzlement, fortunately, does not stop these microscopic problematica from being supremely useful in stratigraphy.

The efficient distribution of microscopic forms over very large areas of the globe, objects with complex morphology and frequently rapid evolution, makes microfossils ideal stratigraphic markers. Spores and pollen, distributed by the wind, may even be used to cross-correlate marine and non-marine strata. Neither are they difficult to find, since they can be expected in most sediments. Almost any sample, properly prepared and treated so as to extract microfossils, can be made to yield useful results.

This has made micropalaeontology highly important to the oil industry, whose wells are not likely to chance upon, or to preserve intact, zonal macrofossils. Microfossils may also be good environmental indicators, and can also give information about the post-depositional history of sediments (see below).

10.2 The unicellular organism and the five kingdoms

Very many microfossils are produced by organisms consisting of only one cell. Most of these cells consist of an outer membrane enclosing the cell matrix or **cytoplasm** and a **nucleus** containing its genetic material. Such cells are called **eukaryotic**, and they are distinguished from the most primitive type of cell, which lacks any true nucleus. Such **prokaryotic** cells, like the bacteria and blue-green algae (see below), make up the first of the five kingdoms of life, the **Kingdom Monera**.

We instinctively think of life as divisible into two kingdoms, not five; we think of the plants and the animals. But at microscopic level such rigid ideas do not hold. Here, many organisms have both plant-like and animal-like characteristics. Some microorganisms, for example, have photosynthetic pigments like plants while at the same time ingesting food particles, like animals. Even the multicellular fungi cannot be properly classed as plants, because they secrete enzymes and digest external food sources, such as wood. They are placed, therefore, in our second kingdom of life, the **Kingdom Fungi**.

The third kingdom is the **Kingdom Protista**, which comprises mostly motile, unicellular organisms. One group within it, the **dinoflagellates** (see below), have whip-like **flagella** for propulsion and contain photosynthetic pigment. But other groups, such as the Foraminifera or **Radiolaria**, ingest food particles, and so are more typically 'animal' in their biology.

The two remaining kingdoms are the true plants (**Kingdom Planta**) and the animals (**Kingdom Animalia**). True plants may give rise to microfossils by the production of spores. True animals may themselves be microscopic (such as the **ostracods**, which belong to the Phylum Arthropoda) or may create microfossils by disintegration.

10.3 Size and scale in single cells

In his book *On Growth and Form* the great biologist D'Arcy Wentworth Thompson said: 'Everywhere nature works true to scale, and everything has its proper size accordingly.' So there are no giants or leprechauns, and no amoebae the size of footballs simply because the physical constraints upon these different anatomies and their component materials make such things impossible.

So why exactly cannot an amoeba be the size of a football? Because there is a disparity between the rate of increase of volume and that of surface area. Anglers know that fish need only increase in length by one inch to double in weight. In a spherical cell of radius R, surface area increases as R^2, whereas volume rises by R^3.

Thus for a unicell that needs its external membrane to take in oxygen and void waste, there comes a point at which the area of this surface is insufficient to meet the demands of the volume. To increase further, the cell must change its shape to one with a larger area per unit volume.

We have already said that foraminifers may grow to many centimetres in diameter — yet they are only single cells. Flattening, which creates shapes with a high ratio of surface to volume, is the rule among such giants.

In this chapter we shall be dealing with organisms much less than a millimetre in size. To describe them, we shall use the **micrometre**, a millionth part of a metre (or a thousandth part of a millimetre). This unit of length is represented by μm. Table 10.1 shows the range of sizes shown by the more common microfossils.

10.4 Kingdom Monera

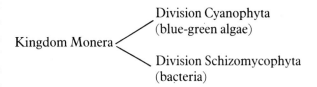

10.4.1 Blue-green algae (Precamb.–Rec.)

Blue-green algae live by photosynthesis, but employ a pigment called **phycocyanin** rather than chlorophyll. This accounts for their distinctive colour. Their cells are minute (25 μm or less), lack nuclei and cannot reproduce sexually. Although they propagate mostly by growth and fragmentation, they do produce a kind of 'spore' by subdividing a single cell.

These are extremely hardy organisms, surviving at extremes of temperature, salinity and desiccation. They are found today in deserts, around hot springs and in hypersaline lagoons where they are usually the *only* living things. Here they can survive, away from competition for space by other algae and free from the depredations of grazing animals.

Blue-greens were the dominant life form on earth for about 2000 million years of the Precambrian, when such competition as now forces them into their inhospitable niches had not evolved. Then as now they grew in laminated algal mats, called **stromatolites** when fossilized (Fig. 10.1). Each lamina may record a daily increment or even a tidal cycle, and the oldest come from the Pongola System of South Africa (3100 my). Single cells, possibly representing blue-green algae, also occur in the Fig Tree Chert, which is of much the same sort of age and also from South Africa.

TABLE 10.1. *Size Ranges of the Microfossil Groups discussed in this chapter*

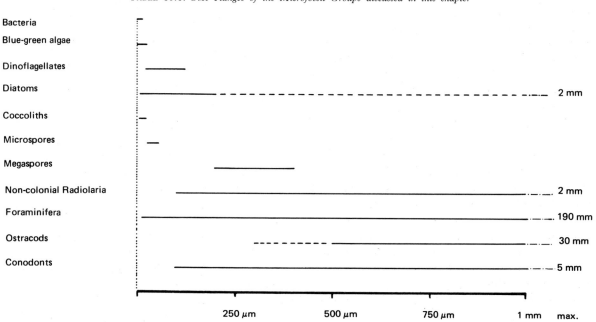

132 *Palaeontology — An Introduction*

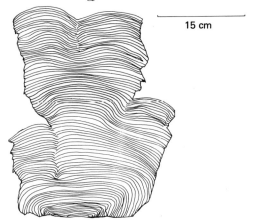

FIGURE 10.1. A columnar, branching stromatolite seen here in longitudinal section. The form of stromatolites is a function of many variables, notably the energy of the environment, sedimentation rate and others. The laminae are created by the entrapment of particles upon the sticky mucilaginous upper surface; this is where the blue-green algae reside, and their upward growth — made necessary by the continuous fouling of the surface — is what builds the structure, layer by layer.

Although today it has been observed that stromatolite shape is governed by environmental forces, a stratigraphy of the late Precambrian and early Phanerozoic has been attempted, using a supposed evolutionary series in stromatolite form. Blue-green algae continue to be of geological importance, especially in limestone-depositing areas and in shallow or partially emergent environments. But their stratigraphic use is strictly limited.

10.5 Kingdom Protista

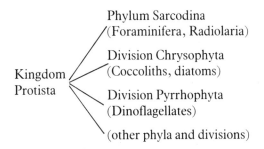

[*Note*: The term 'division' is a botanical taxon equivalent to the zoologist's 'phylum'. Their use together here reflects the old bipartite 'animals-and-plants' tradition under which these groups were studied.]

10.5.1 Dinoflagellates (?Sil.–Rec.)

These are 'borderline' creatures, with both 'plant' and 'animal' characteristics. They seem to have originated in the Palaeozoic — although possibly they are even older — and they are still with us today. In these living forms sexual reproduction is rare, and the manner of cell division during asexual reproduction suggests a link with the prokaryotes. Most dinoflagellates live in the upper waters of the ocean, the **photic zone**, where light can penetrate sufficiently for photosynthesis. Here they may become so abundant as to tinge the water orange or red. Such **red tides** may kill large numbers of fish.

There are two stages in the life cycle: a motile **planktic** stage (Fig. 10.2) and a **cyst** stage. The cyst forms under adverse conditions (such as reduced autumn temperatures) and, being very resistant, is found in the fossil record. There are three basic cyst types (Fig. 10.2). All have an exit pore called the **archaeopyle** which forms by the removal of plates in the apical region. It is via this hole that the organism escapes upon the return of favourable conditions.

Some dinoflagellates inhabit a remarkable habitat — the living tissues of invertebrates. Many invertebrate groups carry symbiotic micro-organisms in their tissues (e.g. the giant clam *Tridacna* and other bivalves have been mentioned — see Chapter 6, III), but the most celebrated examples of this phenomenon are the scleractinian corals (Ch. 8). Their symbionts, the **zooxanthellae**, are often loosely referred to as 'algae', but they are actually dinoflagellates. Certain giant foraminifera also carry such symbiotic dinoflagellates.

Dinoflagellates are widely used as biostratigraphic tools in Mesozoic and Caenozoic rocks, but they are also useful in environmental analysis. Cyst-type, for example, may reflect nearshore or offshore conditions.

10.5.2 Diatoms ((Jur.?) Cret.–Rec.)

Diatoms are non-motile unicells with photosynthetic pigment and a silicified cell wall. They date from the Cretaceous (possibly the Jurassic), occur in many environments and are still living. They contribute to the sediment of the deep sea and may form a lithology known as **diatomite**.

The exterior 'shell' of a diatom is called a **frustule** and consists of two plates (Fig. 10.3 i). One is larger than the other and overlaps it like the two halves of the bacteriologist's Petri dish. There are two basic types of frustule, circular (**centric**) and oval (**pennate**). They are often dotted with tiny holes called **punctae**, whose outer openings are blocked by porous plates (**sieve membranes**).

Diatoms can be found not only in the sea, but in fresh water, the soil and even on the leaves of plants. In the ocean, however, they stand at the bottom of many food chains and are the primary producers in the marine ecosystem. Like dinoflagellates, they populate the photic zone, and being of silica they are preserved in the sediments of the deep sea where calcite would go into solution under the high pressures. Diatomite, however, may also form in lakes.

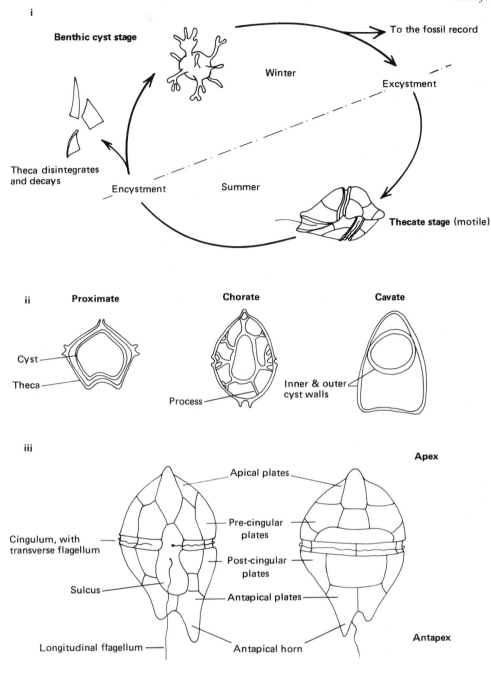

FIGURE 10.2. (i) Dinoflagellate life-cycle. (ii) Types of dinoflagellate benthic cyst. (iii) Morphology and terminology of dinoflagellate thecae.

We shall see below that the fact that calcareous skeletons begin to dissolve at depth sorts out the components in the great rain of potential microfossils falling from the surface waters. Below a certain depth in any ocean, no calcareous fossils can survive. This is the **Calcium Carbonate Compensation Depth (CCCD)**, and its precise depth varies from ocean to ocean, depending upon such factors as the rate of supply of calcareous material from the surface.

Diatoms are used stratigraphically in the Cretaceous and Tertiary for both deep sea and freshwater sediments — though cross-correlation between the two is not easy. Certain living diatoms are known to be sensitive to water temperature, and evidence of palaeo-temperatures may be gained from fossil assemblages. Silica deposits formed by diatoms are commercially quarried in some areas.

10.5.3 Coccoliths (U.Trias.–Rec.)

Coccoliths are the component parts of the armour of the coccolithophore (Fig. 10.3 ii). This is a unicellular, photosynthetic protist which lives mostly in tropical seas. Coccoliths are calcareous and extremely small (3–15 μm), forming when the coccolithophore disaggregates upon death. In rare cases, however,

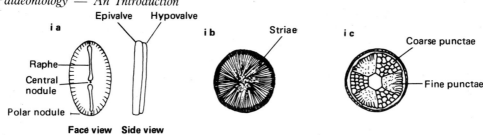

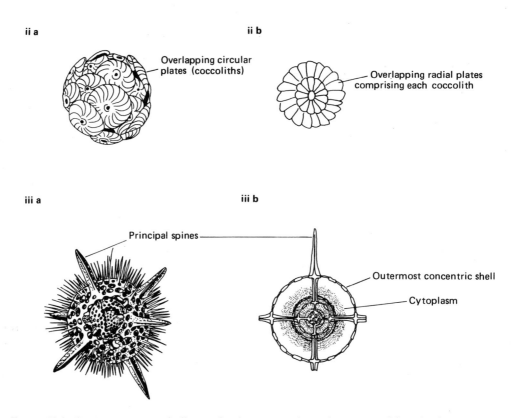

FIGURE 10.3. (i) (a) a pennate (oval) diatom, showing construction and ornament of frustule. (b, c) centric diatoms *Melosira* and *Actinoptychus* showing striations and punctae. (ii) (a) a Recent coccolithophore *Cyclococcolithina*,' showing its construction from many coccoliths (redrawn from Brasier, 1980). (b) a coccolith. (iii) radiolarian *Actinomma asteracanthion*. (a) cutaway drawing to show the principal spines continuous from one lattice shell to another, forming radial beams. Cross-section (b) shows these concentric shells and their relationship with the cell protoplasm (see text): redrawn after Bütschli.

encystment prior to death can preserve the armour in its original intact condition. Such a fossil is called a coccosphere.

Details of coccolith morphology are usually observed with an electron microscope. Being calcareous, they tend not to occur in deep ocean deposits, but in shallower water they can create huge masses of sediment, and have done so since their origin at the end of the Triassic. Their subsequent diversification has made them the principal micropalaeontological zone fossils of the Mesozoic and Caenozoic, and their abundance created thick calcareous oozes over the extensive shelf-sea floor of the Cretaceous and Tertiary.

We know these oozes as **chalk**. Cretaceous chalks comprise about 20% coccoliths, while Tertiary chalks may reach 90%. The recognition of 'warm' and 'cold' coccolith populations may even allow their use in palaeoenvironmental analysis.

10.5.4 Radiolaria (M. Camb.–Rec.)

The radiolaria produce delicate siliceous endoskeletons which average 100–2000 μm in diameter. They are unicellular, marine, and useful both stratigraphically and environmentally.

The skeleton (Fig. 10.3 iii) serves to 'compartmentalize' the functional units of the cell. An innermost chamber houses the nucleus, a middle zone holds the **endoplasm** and an outer shell encases the ectoplasm and the so-called **calymma**. The latter is a kind of ectoplasmic froth containing vacuoles. The radial spines lend rigidity to the perforate shells as well as the **pseudopodia**, cytoplasmic extensions which the organism puts out to entrap food. The food is ingested, and is digested in the vacuoles of the calymma.

Radiolarian form is breathtakingly various. Not all are siliceous; some are made from strontium sulphate, and others of silica plus organic matter. Radiolaria are widely spread in modern seas, but different forms have different preferences as to depth, temperature and chemical composition of the sea water. Some forms with affinity for cold water may live at the surface near to the poles, but may still occur in tropical regions, at great depths.

Radiolaria accumulate in deep oceans, together with diatoms and certain foraminifera (see below). When the surface conditions are optimal, as they are today in the equatorial Pacific, radiolarian ooze can build up to some thickness.

They are therefore of great use in helping to support the notion of sea floor spreading. Old oceanic crust lying near the continental margins of ocean basins bear sediments of greater age than those near the spreading centre. That this was so was demonstrated principally by using radiolaria as stratigraphic markers.

10.5.5 Foraminifera (?Precamb.–Rec.)

Foraminifera (Figs. 10.4, 10.5) construct a shell known as a **test**. It is either made of organic material (**tectin**) or of particles stuck together (**agglutinated**) or of calcium carbonate. There are rare siliceous forms. Primitively, the test comprises one chamber, but in most forms extra chambers are added in spiral arrangement. These chambers remain connected by holes, called **foramina** — and it is these which give the group its name.

The cell matrix, like that of radiolaria, has distinct inner and outer layers. The outer **ectoplasm** lies outside the test, is clear and produces fine, fibre-like pseudopodia for feeding and for locomotion or anchorage. The **endoplasm** is inside the test. It is darker, containing digestive vacuoles and, in some forms, symbiotic dinoflagellates. Food is entrapped by the ectoplasm, and together with its enclosing vacuole is passed to the endoplasm through the **aperture** of the test.

Unlike the protists which we have so far examined, the foraminifera have a fairly complex life-cycle incorporating sexual reproduction. There are two generations (Fig. 10.4), called **schizont** and **gamont**. The schizont is **diploid** (possesses paired sets of chromosomes) and it can multiply asexually. It may also undergo reduction division to form the **haploid** gamont generation. After a while, this form may divide into four motile gametes which, after dispersal, each fuse with another gamete to double up the chromosomes and re-form the diploid schizont.

This is significant for the palaeontologist, for schizont and gamont are not identical.

Basic taxonomy within the foraminifera is based on wall structure. Tectinous walls characterize one group, agglutinated walls another and the calcareous types are subdivided into three by the crystal structure of the calcium carbonate.

But what is the function of the test? It can, for example, be of little use against predators. Fundamentally it seems to serve to subdivide the cytoplasm into functionally differentiated areas. This, together with modifications to shape, helps enable the size increases seen in the group, reaching dimensions unmatched by any other unicellular organism. Surface sculpture on the test may increase buoyancy, and the presence of the test may shield against mechanical damage to the cell, as well as, possibly, the harmful incidence of ultraviolet light.

Some foraminifera live floating at various depths in the water column. But these **planktic** forms only originated in the Cretaceous. Such forms as *Globigerina* today comprise a large quantity of deep-water ooze. And since different forms float at different depths, deep-water deposits will have a greater diversity of planktic forms than shallow sediments. In the deepest water, however, the tests go rapidly into solution, and only agglutinated foraminifera survive in sediments from these depths.

The majority of foraminifers are **benthic**, and these large forms are dominant in shelf-sea sediments where they may even be the major rock-former, e.g. the Nummulitic Limestone characteristic of much Eocene sedimentation in Europe, the Mediterranean and the Middle East.

Since foraminifers are sensitive to temperature, illumination and salinity, particular assemblages may be found to be characteristic of certain environments.

The Foraminifera date back to the early Palaeozoic and may well have arisen in the Precambrian. Organic-walled types probably gave rise to both the agglutinated and calcareous varieties. Since the Devonian, many new developments and diversifications (notably the arrival of planktic types in the Cretaceous) have taken place, and their plastic and complex form has made them invaluable stratigraphic tools.

10.6 Kingdom Planta — spores and pollen (Dev.–Rec.)

10.6.1 The alternation of generations and the rise of land plants

The spores and pollen of land plants are extremely tough and fossilize readily. Moreover, their dispersal in the wind has ensured that since the Devonian, when

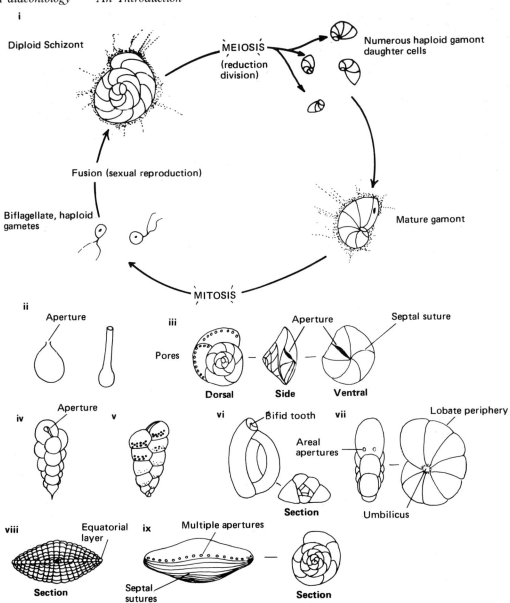

FIGURE 10.4. Foraminiferal life cycle (top) and coiling modes, (ii) Unilocular tests (consisting of one chamber only), (iii) low trochospiral test, (iv) high trochospiral, triserial test (three longitudinal rows of chambers), (v) high trochospiral, biserial test (two rows of chambers), (vi) milioline winding, (vii) planispiral lenticular test, (viii) Annular complex type, (ix) planispiral fusiform coiling.

land flora first became significant, these tiny objects have been incorporated into most sedimentary rocks. But to use them wisely it is essential that their function is understood, and this is part and parcel with those evolutionary developments which led to the conquest of land by the Kingdom Planta (Ch. 11).

The plant life cycle involves two generations. One reproduces asexually by the production of millions of wind-borne spores (and is hence known as the **sporophyte**). The other results from the germination of these spores. Upon this form, male and female reproductive organs are situated, and fertilization of the female leads to the growth of a new sporophyte. This sexual generation is known as the **gametophyte** (Fig. 10.6 i, ii).

For the male sperm to reach the female organ, the gametophyte must be covered in a thin film of moisture in which the sperm can swim. So it is that plants with this system of reproduction are confined to damp places — like the mosses and ferns. Ferns have gone some of the way towards reducing the importance of the gametophyte, for it is small and insignificant when compared with the sporophyte, which is the plant familiar to all of us. Contrast this with the case of the mosses, in which the sporophyte generation is merely a 'parasitic' outgrowth upon the gametophyte.

In mosses the spores which are released by the sporophyte and which give rise to the gametophyte generation are all the same. This is because the moss

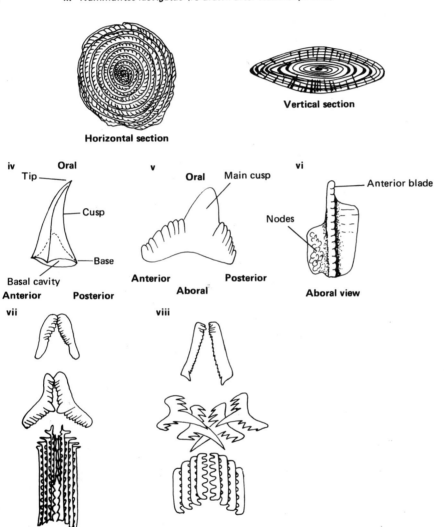

FIGURE 10.5. Foraminifera (i)–(iii) and Conodonts, (iv) a simple cone, (v) a blade, (vi) a platform element. Note anterior/posterior, oral/aboral terminology. (vii) *Scottognathus*, a conodont 'apparatus'. (viii) *Illinella* another apparatus, employing different elements.

gametophyte possesses both male and female organs on the same plant (Fig. 10.6 i). In ferns, on the other hand, two sorts of spore are produced; the smaller **microspore** gives rise to a gametophyte with male organs, while the larger **megaspore** forms a female gametophyte (Fig. 10.6 ii). These two types of life cycle are called **homospory** (spores all the same) and **heterospory** (microspores and megaspores). But neither is by any means perfectly adapted to life on land.

Clearly, the dependence of the gametophyte upon moisture means that its separate existence as a free-living generation prevented the more independent sporophyte generation from achieving its full potential as a land plant. It was the seed plants which finally took the decisive step and broke the link with damp places.

In their system (Fig. 10.6 iii) the female gametophyte remains captive on the sporophyte, where it develops within a structure called the **ovule**. This is fertilized by microspores (known as **pollen**). These, upon reaching the female **stigma**, grow a long tube (the **pollen tube**) which extends down through the tissue of the female organ towards the ovule. This pollen tube may be regarded as the last remnant of the male gametophyte. When fertilization occurs, the new sporophyte is produced, but it takes the form of an embryo, or **seed**. Seeds are necessary in this new life cycle, because the captivity of the female gametophyte has removed the dispersal role from the spore stage. The new sporophyte is therefore 'packaged' in miniature, and sent on its way under a variety of ingenious **dispersal mechanisms**.

Naked seeds, lacking a tough outer coat, are produced by the **gymnosperms**, whose name means 'naked seed' and which include the coniferous trees. Flowering seed plants are called **angiosperms**. Their seeds have a resistant outer jacket developed from the covering of the ovule.

10.6.2 Typical spore and pollen morphology

Megaspores may be as large as 400 μm in diameter — nearly half a millimetre — and so are sometimes

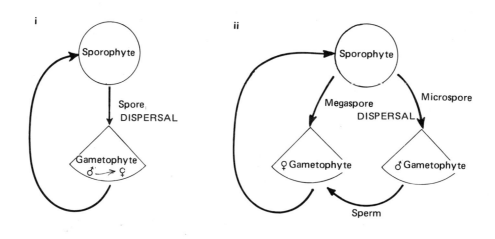

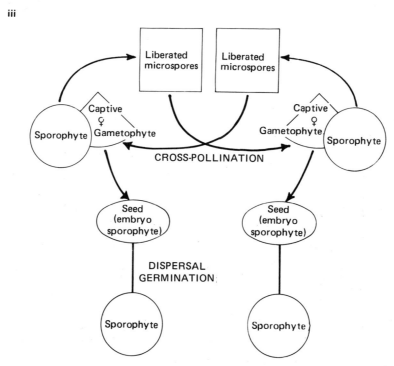

FIGURE 10.6. Alternation of generations in typical plant life-cycles (see text) (i) Homospory, as seen in mosses. (ii) Heterospory, as seen in ferns. (iii) the life cycle of seed plants.

visible to the naked eye. Microspores tend to be a lot smaller — somewhere between 5 and 50 μm. In form and ornament, both types are very varied, although most possess some basic features which are a product of their formative cell-divisions (Fig. 10.7).

Every group of four spores (**tetrad**) had a single **mother cell**. If this cell divided into four all at once, then the arrangement of spores in the tetrad is **tetrahedral**. However, it is possible for the mother cell to split firstly into two, and then for each daughter cell to split again. In this case the four daughter cells (each one of which will form a spore) are arranged vertically, in the same sort of pattern as the segments of an orange.

Each method of division leaves tell-tale traces on the inward-facing surfaces of the spores. Tetrahedral tetrads (the sort which form by simultaneous division) create spores with tri-faceted proximal surfaces, these

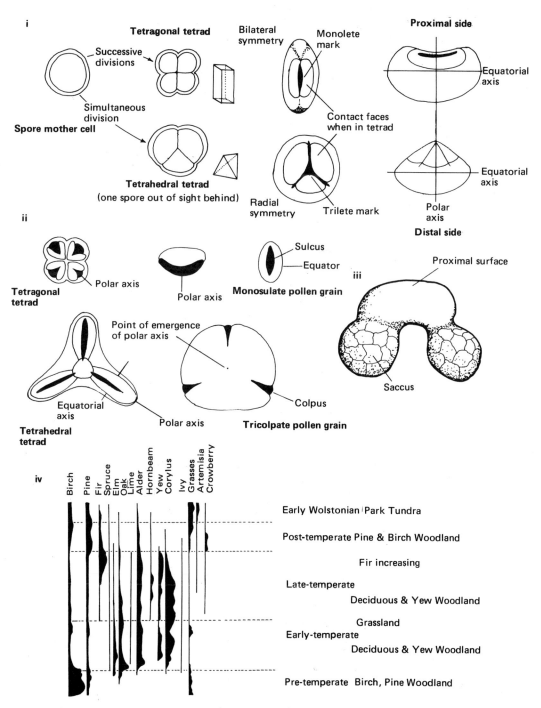

FIGURE 10.7. (i) Morphology of monolete and trilete spores and their formation from a single mother cell. (ii) Morphology of monosulcate and tricolpate pollen and their formation. (iii) Pollen of *Pinus* (pine) showing air-filled sacs. (iv) A pollen analysis of interglacial sediments from Marks Tey in Essex. The relative proportions of different pollens at different levels indicates the floral composition of the time and hence the climate. (Simplified after Turner.)

faces meeting at three sutures ranged at 120° to one another. These three radiating seams are together called the **trilete mark**, and spores bearing them are known as **trilete spores**. Spores which form by sequential division have bi-faceted proximal surfaces, their two faces meeting along a single seam, the **monolete mark**. (Rare, scarless spores are known as **alete**.)

Spores also have a **germinal aperture**. The type and position of this hole, which allows the spore to germinate upon settling, are useful characters for the taxonomist. Also useful are the external sculpture and the presence or absence of **flanges** and **air sacs**.

Pollen grains are also formed by the division of mother cells, and so, as with spores, the grain's basic features bear a relationship to the manner of this division. Unlike spores, however, pollen grain scars are formed on the external faces, not the internal ones.

Monosulcate pollen bears a single groove on the distal edge (Fig. 10.7 ii) and, like monolete spores, form by successive division of the mother cell. Such grains are typical of gymnosperms and grasses.

Tricolpate pollen is typical of flowering plants (other than grasses), and bears three grooves (set at 120°), just like trilete spores. In pollen, however, these grooves serve as germinal apertures, and in some cases may be reduced to the form of simple pores.

Spores and pollen have double walls, the inner one being called the **intine**, and the outer the **exine**. The exine may be deeply sculptured, and may even be inflated into buoyancy sacs to aid dispersal (see above). Spores with such air sacs are called **cavate**, and pollen similarly endowed is called **saccate** (Fig. 10.7 iii).

Exine is extremely tough, serving to protect the protoplasm within, but its resistance also permits ready preservation and allows the palynologist to isolate material using strong acids such as hydrofluoric (HF) to remove the rock surrounding the spores. These are then found among whatever survives this acid digestion, which is collectively known as the **Acid-Insoluble Residue (AIR)**.

10.6.3 Geological and stratigraphic value

Spores are employed to great effect over the entire period from the Devonian to the present. This makes the selection of specific examples rather difficult. Suffice it to say that palynology can date otherwise unfossiliferous sediments, such as continental clastics, and their use in the cross-correlation of lake and river sediments with established marine sequences is unrivalled. Their widespread utility has made palynology a subject much favoured by the oil industry, whose interest has largely fuelled its development over the last 30 years.

Another oil-related application of spore material is in the estimation of **sediment maturity**. To produce useful hydrocarbons, organic-rich sediments need to undergo a period of 'cooking' or **maturation**. The duration and temperature of this process largely determines the nature of the product, i.e. whether oil, condensate or gas. By examining the depth of colour in spore exines, a measure of this may be made.

Environmentally, pollen is of great importance in describing changes of tree population with time. Such changes may be brought about by climatic fluctuation or by the activities of man, and they are recorded in bog or lake sediments as changes in the pollen assemblage.

From the interglacial sediments of Marks Tey in Essex, the pollen analysis reveals four major climatic regimes reflected in the tree population. A pine and birch assemblage with a lot of grass pollen is interpreted in terms of a pre-temperate climate shortly following the retreat of the ice. It is succeeded by assemblages which mark the appearance of warmth-loving trees such as oak, hazel, alder and yew. The sequence is completed with the gradual return of pine, birch and grass which in turn anticipate the re-advance of the ice sheets in the succeeding glacial period (Fig. 10.7 iv).

10.7 Kingdom Animalia

10.7.1 Ostracods (Camb.–Rec.)

Phylum Arthropoda   Class Crustacea
Subclass Ostracoda

Ostracods are shrimp-like organisms which have enclosed themselves in a pair of hinged **valves**. They are typically one millimetre to a few millimetres in length, and inhabit seas, fresh water and even the soil. Most, however, are benthic or free-swimming marine. There is a wide range of feeding habit, the sexes are separate, and the male ostracod is responsible for producing the largest spermatozoon in the entire animal kingdom.

Palaeontologists are only concerned with the valves, for the rest of the animal (Fig. 10.8) is rarely preserved. These valves are made of calcified chitin and are joined dorsally by an uncalcified **ligament**. The valves are closed by **adductor muscles** clustered centrally and may articulate by **teeth** and **sockets**. The similarity to bivalved molluscs in terms of form and function is striking, but it does not extend to the valve edges. In ostracods, one valve, being slightly larger than the other, commonly overlaps it along the ventral margin.

There are no growth lines on ostracod valves because, being arthropods, the valves are regularly moulted along with the rest of the exoskeleton. There is also an internal structure formed by the inward-folding of the ventral edge, called the **internal lamella**. This is often used in taxonomic description, along

Microfossils 141

i Ostracod anatomy (*Skogsbergia*, ♀) (After Claus from Calman)

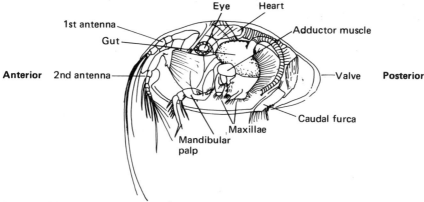

ii Internal features of the carapace (redrawn from Markhoven)

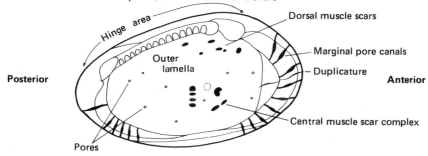

iii Section of outer portion of valve, showing outer lamella and duplicature (modified from Kesling)

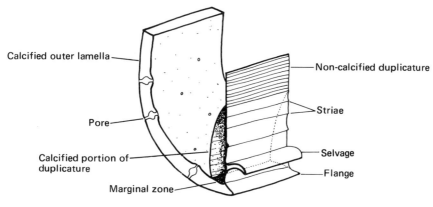

FIGURE 10.8.

with such features as the type of ornament, the hinge mechanism, the position of the muscle scars and the type of sexual dimorphism (see below).

Various kinds of pore perforate the carapace to allow the passage of sensory bristles or **setae**. This tactile system may be supplemented by vision, in which case the valves may contain a transparent 'window' directly over the position of the eye. The fact that ostracods moult, like trilobites, means that for most forms there may be many forms of carapace, each reflecting a discrete stage in the animal's development (Fig. 10.9). Further complications involve sexual dimorphism (Fig. 10.9), which may result in there being **brood pouches** on the female and a retention of juvenile form (or the assumption of a more elongate shape) by males.

Ostracods are useful stratigraphically, especially in Mesozoic and Caenozoic non-marine sediments. But their greatest success has been as tools in the interpretation of ancient environments. The nature of the population, such as its abundance and diversity, or the ratio of males to females, can be very informative. There are also specific features of the carapace, such as thickness or degree of calcification, which may be significant. Freshwater ostracods are lightly calcified and smooth, whereas benthic marine types are thick-shelled and may be ornamented with ribs and tubercles. In fine sediments, snowshoe weight distributors may be attached to the carapace, and burrowers, like burrowing bivalves, tend to be smooth and extended.

Ostracods are the second most abundant modern

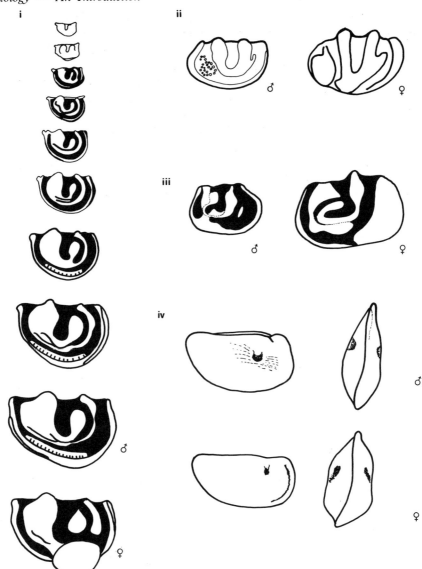

FIGURE 10.9. (i) Ontogenetic series of moulted carapaces, and mature male and female specimens of *Craspedobolbina*. (ii) Sexual dimorphism in *Zygobolbina*. (iii) Sexual dimorphism in *Mastigobolbina*. (iv) Sexual dimorphism in *Sansabella*.

microfaunal group (after the foraminifers) and, especially in brackish or freshwater conditions, may be so plentiful as to become rock-formers; e.g. the Cypris Freestone, in the Purbeck (U. Jur.) of Dorset.

10.7.2 Conodonts (U. Precamb.–U. Trias.)

Conodonts have been studied since the 1850s, yet they remain the most enigmatic of fossils. They are tooth-like, spiky objects about 0.1–5 mm in length, composed of apatite (calcium phosphate), and are quite abundant. Their resistant nature results in excellent preservation, and with their diverse and complex form they provide plenty of opportunity for detailed description. Moreover, once they have been extracted from the rock with acetic or formic acid, they can be employed as excellent stratigraphic markers.

There are three basic types of conodont element (Fig. 10.5), **cones, bars/blades**, and **platforms**. Their growth was incremental, which suggests that they were originally covered with living tissue. They have a finely laminated microstructure. Basally there is a cavity, and in this region the denticle may flare out to form a flange-like **basal body**. Conodonts often come in left- and right-hand forms, mirror images of each other, suggesting that originally they worked in opposition.

In fact, many conodonts have been found to compose **conodont apparatuses** (Fig. 10.5). These are preservations of up to twenty-two elements in their 'life position', i.e. in the positions relative to one another that they adopted when they were incorporated in some soft-bodied creature of which we know nothing.

Clearly, any speculations as to the original function of the elements, or the nature of the animal which made them, should take account of these remarkable assemblages. The likelihood is that they acted as a straining or filtering system surrounding a portion of the gut of the elusive 'conodont animal'. They were probably not teeth, since it is likely that they were covered completely in living tissue. Their pristine, unworn condition also militates against this possibility. As yet, few convincing solutions to the problem have been produced, very few 'conodont animals' have been seriously suggested, and none of these accounts for the apparatuses.

Despite these problems, conodonts do at least serve some useful purpose as stratigraphic tools, especially in the Palaeozoic. In the Tremadoc, the resolution they provide equals that of graptolites and even surpasses them in the Arenig. It has also been found recently that their colour may be used, like that of spore exine, to assess the maturity of potential hydrocarbon sources.

11.
Vascular Plants

11.1 Introduction

Vascular plants (**tracheophytes**) have a system of tubes inside their stems which conducts water from their underground parts to the rest of the organism. Their origin, probably during the Silurian, was therefore intimately associated with the move on to the land, and subsequent developments can be viewed as further adaptations to the terrestrial habit. In this sense, therefore, the approach in this chapter is the same as that adopted in Chapter 8 with respect to the vertebrates.

Vascular plants were much more preservable than anything which had gone before, yet the move to land habitats which accompanied their development meant that their fossil representation remained sparse. Despite this, however, there are many 'windows' in the palaeobotanical record which afford some tantalizing glimpses into plant history and provide a stimulus to new hypotheses. In this account we shall try to describe the major groups in their geological and evolutionary context, but to make sense of the progression, some knowledge of basic botany is an advantage. The reader is therefore strongly advised to peruse subsection 10.6.1, where many concepts and terms relating to plant life-cycles are explained.

11.2 Palaeozoic plants without seeds

In most major trachaeophyte groups we have both living and fossil material to help us. So it is with the earliest vascular plants, the **psilophytes**.

11.2.1 Psilophytes (U. Sil.–Rec.)

Despite the recovery of spores bearing trilete marks from older deposits, the first undoubtedly vascular plants occur in rocks of the Upper Silurian. These are vascular plants at the barest possible level of complexity, with no true roots. Aerial shoots, lacking leaves, rise from a horizontal subsurface stem called a **rhizome**. These shoots may be seen to branch by simple splitting into two (**dichotomy**) or three (**trichotomy**), and at their ends they bear homosporous **sporangia**. The earliest psilophyte is found in Downtonian rocks in Herefordshire, Pembrokeshire (U.K.), Czechoslovakia, Podolia (U.S.S.R.) and North America. Assigned the genus *Cooksonia*, it grew to about 7 cm and was leafless, with ovate to spherical sporangia (Fig. 11.1 i).

Better known (and much better preserved) are the plants of the Rhynie Cherts (M. Dev.) in Aberdeenshire, Scotland. First specimens were spotted in stone blocks in a wall, and palaeobotanists Kidston and Lang eagerly traced the blocks to their source. There they found one of the most important fossil localities in the world. Silicification had preserved the plants in the utmost detail (Fig. 11.1 ii), even down to the cellular level.

Rhynia was a similar plant to *Cooksonia*, but larger — reaching some 20 cm in height. More complex was *Asteroxylon*, growing to perhaps half a metre and having its upright stem clothed in leaf-like scales (Fig. 11.1 iii). Branching was by the formation of offshoots from a dominant axis (**monopodial**), and some of these side shoots were bare of leaves and bore sporangia. Similarly specialized reproductive organs were found in *Zosterphyllum* (Fig. 11.1 v), with conspicuously clustered sporangia.

In addition to the psilophytes, there is a wide array of problematic fossils which are found in late Silurian and early Devonian sediments, plants which cannot be fitted into any of the recognized major groups. Nevertheless, many seem to show 'tendencies' towards a certain condition or another, and it may be that this group of 'homeless' plants with no fixed taxonomic abode includes the ancestors of several great plant dynasties.

One such plant, with the eye-catching name of *Enigmophyton superbum* (Fig. 11.1 iv), had conspicuous leaves, and although its sporangia bore similarities to the **lycopods** (see below), its true affinities remain undecided.

11.2.2 Lycopsida (U. Sil.–Rec.)

By comparing the three living and two extinct orders of the Lycopsida, it is plain that the modern club-mosses are a pale reflection of the ancient group. Most lycopsids were heterosporous and bore small

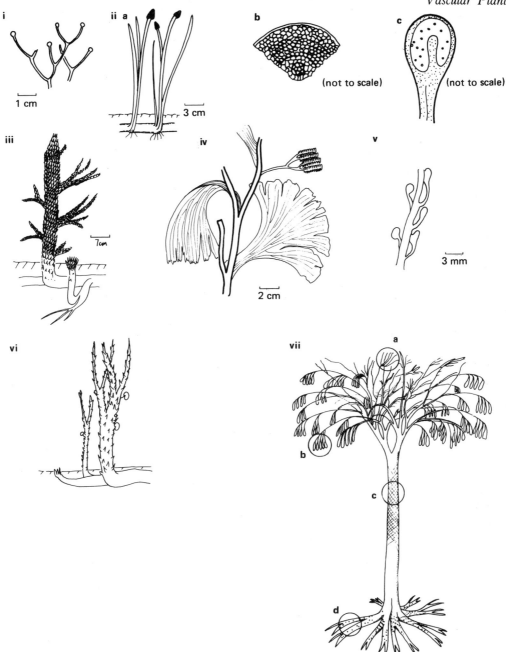

FIGURE 11.1. (i) *Cooksonia* sp. (ii) *Rhynia gwynne-vaughani* (a) restoration (b) transverse section of stem (c) longitudinal section of terminal sporangium. a and c after Kidston and Lang; b after Andrews. (iii) Restoration of *Asteroxylon mackiei* (after Kidston and Lang). (iv) *Enigmophyton superbum* (redrawn from Høeg). (v) Axis of *Zosterophyllum* with sporangia. (vi) *Drepanophycus spinaeformis* (after Krausel and Weyland). (vii) Diagrammatic reconstruction of *Lepidodendron*. These tree-like or arborescent lycopods are known to have exceeded 35 m in height, with trunks reaching 1 m in diameter basally. Different component parts are ringed. (a) twigs (b) cones (c) main trunk and (d) roots. Many of these are assigned to their own so-called 'form genera'; e.g. the cones are referred to the genus *Lepidostrobus* and the roots to *Stigmaria*.

leaves, their branching was a combination of the dichotomous and the monopodial, and spores were formed in **cones**. These are clusters of fertile, scaly leaves, called **sporophylls**.

Baragwanathia is the oldest known specimen. It comes from Australia and has actually been known to occur obligingly on the same slab as the zone monograptid! In the Devonian, such forms as *Drepa-nophycus* (Fig. 11.1 vi) and *Protolepidodendron* occur, the latter in the M. Devonian of Germany. It had stems up to 6 mm broad and forked leaves up to 3.5 mm long.

The Order Lepidodendrales is perhaps the most important of all. In Coal Measure time (U. Carb.), *Lepidodendron* (Fig. 11.1 vii) grew to over 35 m, with a trunk 1 m across. This thickening was partly achieved

146 *Palaeontology — An Introduction*

by **secondary growth**, producing woody material (and in modern trees, this secondary thickening produces the familiar annual rings). But most of the thick trunk of *Lepidodendron* was still composed of spongy tissue.

At the top of the towering trunk was a crown of branchlets, mostly leafy, some specialized to bear cones. At its foot there was an unusual branching root system known as the **stigmarian base**. Upon these roots, spirally arranged rootlets were carried, in the same kind of arrangement as the leaves on the stem. These leaves, when shed, left characteristic scars (Fig. 11.2) which are useful in identification.

Because plants are rarely preserved whole, their component parts (whose unity of origin is often not realized at first) are named under what are called **form genera**. So it is that the root of *Lepidodendron* is called *Stigmaria*, its leaves *Lepidophylloides* and its cones *Lepidostrobus* (Fig. 11.2 iv).

The giant lycopsids flourished in the deltaic swamplands of the Carboniferous, and are presumably a

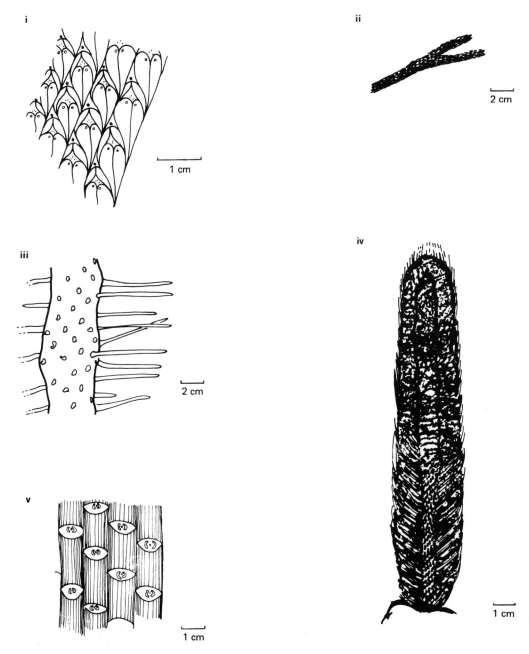

FIGURE 11.2. Carboniferous lycopods. (i) Exterior of large *Lepidodendron* branch or stem showing typical leaf-scar pattern. (ii) Small twigs of *Lepidodendron* invested with scaly leaves. (iii) *Stigmaria* showing lateral roots and scars. Note spiral arrangement. [Note also that scales are tentative in these diagrams. The size of a trunk or a root may vary considerably]. (iv) *Lepidostrobus*, the cone or 'strobilus' of *Lepidodendron*. (v) Exterior of another lycopod, *Sigillaria*: portion of branch or stem showing typical leaf-scar pattern. (All diagrams from specimens or photographs.)

major component of our coal. They died out at the end of the Palaeozoic.

11.2.3 Sphenopsida (L. Dev.–Rec.)

The sphenopsids (Fig. 11.3) are a distinctive group of jointed plants which were very important in the Carboniferous and which survive today in one genus, the horsetail *Equisetum* — a common inhabitant of hedgerows, boggy ground and waste places. These plants are all characterized by a rhizome system from which arise the vertical stems with their radiating appendages in whorls at each joint or **node.** In *Equisetum* these appendages are actually stems, not leaves, and are themselves jointed. Where they meet the main stem a collar of very reduced true leaves may

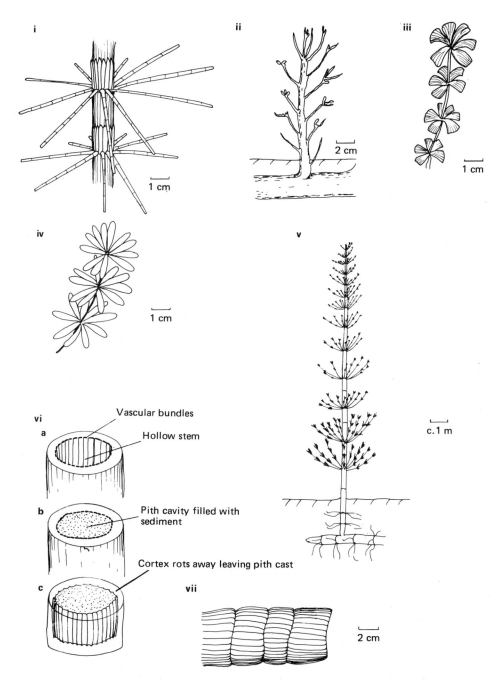

FIGURE 11.3. Sphenopsida. (i) Modern horsetail *Equisetum*, showing jointed main stems, radial stems and vestigial leaves as collar around axis. (ii) Restoration of *Protohyenia* (after Ananiev). (iii) *Sphenophyllum* — a Carboniferous herbaceous sphenopsid. (iv) *Annularia*, a common Coal Measure form. (v) Reconstruction of the Carboniferous tree *Calamites*, of whose foliage *Annularia* is the form-genus. These giants, like the modern horsetail, grew from underground stems or rhizomes. (vi) Mode of preservation. The centre of calamite stems was occupied by a pith cavity. This became filled with sediment, and the form preserved (vi c and vii) is therefore a 'pith-cast' bearing the impression of the internal surface of the hollow stem.

be seen. The stems are hollow, grow to about a metre tall and bear **terminal sporangia.**

It is interesting to note that although the spores of *Equisetum* are all the same size, some produce male and some female gametophytes. In other words, there is functional heterospory in a homosporous plant.

The earliest sphenopsid was *Protohyenia* (L. Dev., Fig. 11.3 ii), which had a rhizome and branched aerial shoots, though only vague whorling of offshoots. *Sphenophyllum* (Carb., Fig. 11.3 iii) was a scrambling or climbing plant with delicate stems and fan-shaped leaves. *Calamites*, most important in the U. Carboniferous, was like a monster *Equisetum*, growing to 25 m from a massive rhizome and bearing shoots up to 35 cm in diameter. Older stems were quite woody, with secondary thickening, but the large central **pith cavity** was often preserved as a cast as a result of sediment infill (Fig. 11.3 vii). These, and the radiating leaves (assigned the form genus *Annularia* — Fig. 11.3 iv), are common fossils in the Coal Measures.

11.2.4 The origin of true leaves

In the above groups, leaves, if present, have been small and scale-like. With the ferns and their related groups we see the first appearance of really large leaves, which probably originated in the following manner (Fig. 11.4).

Firstly, shoots became differentiated into a central axis with regular offshoots. This structure then became flattened into a plane (e.g. *Telangium*, Carb.) and subsequently webbed to form a leaf (e.g. *Rhacopteris*, Carb.).

The main stem of a fern frond is called the **rachis**, and its first and second order branches are called **primary** and **secondary pinnae**, continuing to the smallest subdivision, which are known as **pinnules** (Fig. 11.4 ii).

11.2.5 The fern complex

By about the middle of the Devonian, the earliest fern-like plants had evolved. They were the **pre-ferns**, and they probably gave rise to the two other groups in the 'fern complex', the **true ferns** and the **seed ferns**. These are among the most important groups of plants both in terms of sheer abundance and evolutionary position.

Difficulty arises, however, from the fact that it can be impossible to tell the leaf of a seed fern from that of a true fern. Indeed, the form genera erected to contain typical fern fronds have been found to encompass similar fronds from representatives of both groups. As we shall see, the true ferns and seed ferns may only be reliably distinguished on their reproductive structures.

11.2.6 Filicopsida (Dev.–Rec.)

These are the true ferns. Their frond-like leaves grow from a central crown which may be close to the ground or raised upon a trunk many metres in height (**tree ferns**). The plant is anchored by fibrous roots. Most ferns are heterosporous, the spores originating in small sporangia on the undersides of pinnae. The gametophyte which these spores give rise to is insignificant, a butterfly-shaped pad of cells only a few millimetres across.

True ferns are exceedingly diverse and complex plants. They were significant components of the Mesozoic flora, but in the Carboniferous they appear to have been less important than their evolutionary offshoot, the seed ferns (see below).

11.2.7 Plant evolution reflected in Devonian spore assemblages

The Devonian saw the first of three great revolutions in the history of plant life (Table 11.1). At its inception, nearly all trachaeophytes were homosporous. By its close, heterospory was common, and from it the first rudimentary seed plants had developed. The earliest seeds so far discovered come from the U. Devonian of North America — though rather closer to home Old Red Sandstone sediments at Tongwynlais (near Cardiff) have yielded others.

11.3 Seed plants of the pre-Permian

11.3.1 Pteridospermales (U. Dev.–Jur.)

These are the seed ferns, plants whose separate existence from the true ferns was recognized in the 1900s, when what had hitherto been thought of as filicopsid foliage was found in association with unique seeds (*Lyginopteris* leaves and *Lagenostoma* seeds — Fig. 11.4 iii–vi).

This was nothing less than a revelation, and since that time much of the fern-like foliage known from the Carboniferous has proved to be of pteridosperm origin. The seeds were borne upon the fronds either laterally or apically, but always singly — never clustered into cones. The microsporangia, however, were clustered, but they did not occur on the undersides of the pinnae, as they do in the true ferns.

Some seed ferns were arborescent in habit, others creeping or herbaceous, and the larger ones had considerable secondary wood within their trunks.

11.3.2 Cordaitales (Carb.–?Trias.)

These were magnificent trees up to 30 m tall, with dense crowns of branches bearing strap-like leaves often a metre or more long (Fig. 11.6 i, ii). They produced dense secondary wood, had an extensive

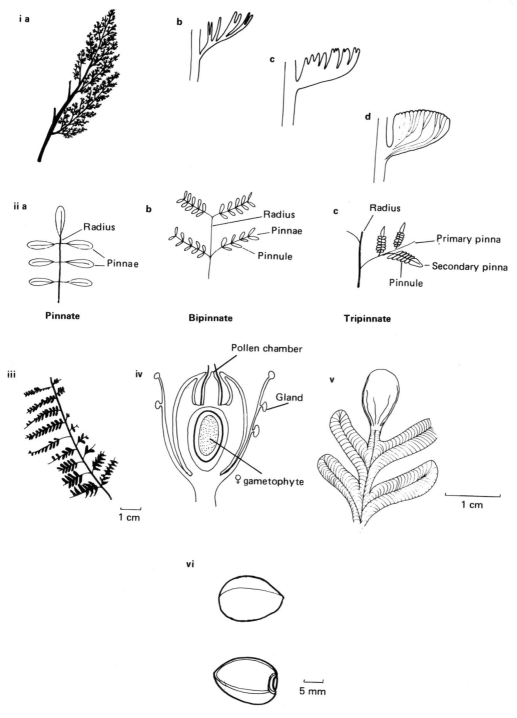

FIGURE 11.4. (i) The evolution of the true leaf according to the 'telome concept'. (a) stem of *Telangium affine* showing densely branched form across which webbing may be envisaged to form a leaf. (b) *Rhacopteris* pinnule (*R. inaequilatera*). (c) Pinnule of *Rhacopteris flabellata* showing less deeply-divided form. (d) Pinnule of *R. lindseaeformis* with divisions eliminated. (From Andrews, 1961). (ii) Compound leaves. (iii) *Lyginopteris* frond. This is a seed-fern. (iv) *Lagenostoma* seed, borne on *Lyginopteris* fronds, in longitudinal section (after Walton, 1940). (v) Seed attached to *Alethopteris* foliage. (vi) Pteridosperm seed with characteristic pointed apex and basal scar where it was once attached to the frond as in (v).

TABLE 11.1. *Broad pattern of plant history, showing three major 'revolutions' in plant-design (shaded areas). See text for details*

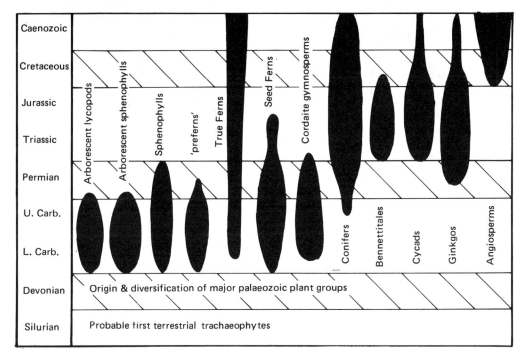

root system and probably inhabited rather higher, drier ground than the typical lycopsids, sphenopsids and pteridosperms of the Carboniferous deltaic swamps. As a result, their remains tend to be drifted and transported. But many specimens are very large, despite this; one incomplete trunk found near Newcastle was 25 m long.

Reproductive structures were borne in the axils of the leaves, and the seeds in cone-like clusters.

11.3.3 Coniferales (U. Carb.–Rec.)

Modern conifers are generally tall trees, each individual bearing male and female reproductive structures. The pollen is often characteristically invested with wings or air sacs to aid dispersal (Ch. 10) and the leaves are typically needle-like and spirally arranged. It is to this group that the largest of all trees, either living or fossil, belongs, the giant redwood of California (*Sequoia*) which may reach 100 m in height. Also familiar is the *Araucaria* or 'monkey-puzzle', widespread in ornamental use since the last century. But it is in forms such as the pine (*Pinus*) and larch (*Larix*, a deciduous form) that conifers dominate boreal and high-altitude forests today.

Palaeozoic conifers such as *Lebachia* and *Ernestiodendron* (Fig. 11.6 ii, iv) were but small trees, yet they have the characteristic type and arrangement of leaves seen in all conifers, and their cones were of similar (though not identical) structure.

It was in the Mesozoic that conifers became of real importance for the first time. The Wealden Series (L. Cret.) contains many fossil conifers, and many of the silicified trees of the Painted Desert (Triassic, U.S.A.) were also coniferous, e.g. *Araucarioxylon*, a form which may have reached 40 m.

By the end of the Mesozoic, however, coniferous plants were beginning to give ground (literally) before the flowering plants.

11.4 Non-flowering seed plants of the Permian and post-Permian

The second big change in the earth's vegetation took place in the Permian (Table 11.1). Many new seed plants originated, later to become major factors in the Mesozoic flora.

11.4.1 Ginkgoales (Perm.–Rec.)

There is but one surviving species of this group, *Ginkgo biloba* (Fig. 11.6 v), a true 'living fossil', saved from extinction (so the story goes) by cultivation over the millennia in the monastery gardens of the Far East. Now it is widely planted in parks and gardens as an ornamental.

It grows to about 30 m in height and has a broadly conical shape when young, filling out when older. Its branching pattern is rather haphazard, but its very distinctive fan-shaped leaves (which are shed in winter) grow on short stalks. These leaves are fairly common as fossils, and occur in the Deltaic Series of Yorkshire (Jurassic) as well as the Ardtun Leaf Beds (Tertiary) of the Isle of Mull. These ancient ginkgos may have deeply indented, palmate leaves, suggesting that many more species then existed.

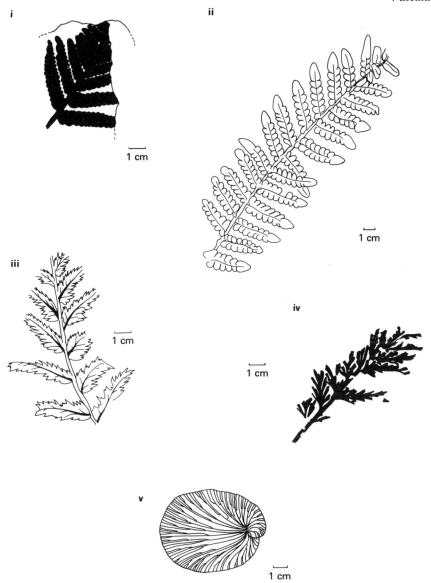

FIGURE 11.5. Fern-like foliage from the Carboniferous. (i) *Asterotheca* sp., Gilfach Goch, Glamorgan. (ii) *Neuropteris scheuchzeri*. Danygraig, Swansea. (iii) *Mariopteris* sp. Maesteg, Glamorgan. (iv) *Alethopteris lonchitica*. Forest of Dean, Gloucestershire. (v) *Cyclopteris* pinnule, Coalbrookdale, Shropshire. (Redrawn from photographs.)

Ginkgo is a unisexual plant. The male form tends to be preferred in gardens because the fruits (which, of course, occur only on female trees) have a strikingly rancid, sweaty odour.

11.4.2 Cycadales (Trias.–Rec.) and Bennettitales (Trias.–U. Cret.)

As with ferns and seed ferns, cycads and members of the Bennettitales may appear very similar to one another. But this superficial likeness masks more fundamental divisions. Certain macroscopic features can be used to distinguish them (see below), but only in those rare specimens where they can be seen. In the usual specimen, differentiation is difficult, if not impossible.

Cycads of the modern world grow very slowly. They have a crown of frond-like leaves with two rows of pinnae arranged on a single stem. These fronds may be metres long and, being tough, are not uncommon as fossils. The crown usually sits on top of a trunk which is clothed in the bases of dead leaves and packed out with pithy material. The trunk may branch or it may not, and in extremely old specimens can grow to 20 m.

The cones of a cycad are borne singly in the centre of the crown, and the plants are exclusively unisexual. In the Bennettitales, however, cones occur in clusters in among the leaf bases and bear male and female organs on the same structure.

Figure 11.7 shows *Williamsonia*, a bennettitalean from the Jurassic of India. Examples of this genus also

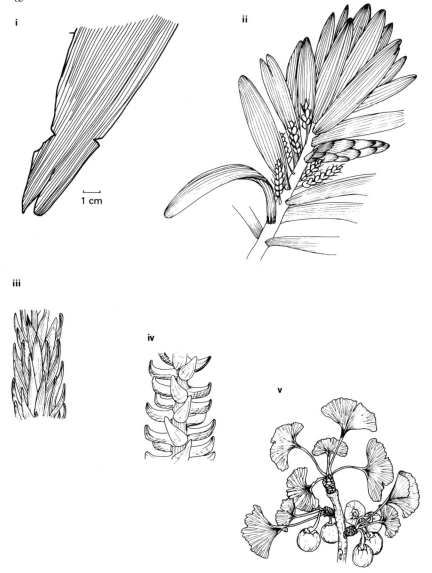

FIGURE 11.6. (i) Part of a strap-like leaf of the Carboniferous gymnosperm *Cordaites principalis* (Caerphilly, Glamorgan). These leaves may have grown to over 1 m in length and were borne on branches atop trunks well over 35 metres tall in many cases. (ii) Reconstruction of a cordaitean branch showing inflorescences in the leaf axils (after Grand d'Eury, 1877). (iii) and (iv) Permian conifers *Lebachia* and *Ernestiodendron* respectively. (v) *Ginkgo biloba*. Note unusual leaves borne on short side-branches. Fruits betray this as a female specimen.

occur in rocks of similar age at Whitby, Yorkshire. It may be useful to point out that despite a vague similarity, neither cycads not bennettitaleans are 'palms'. Palms are flowering seed plants, more closely related to grasses than either of these groups.

11.5 Flowering plants

The third and final revolution in the plant kingdom came in the Cretaceous (possibly beginning somewhat earlier) and completely transformed the surface of the earth. Its consequences were profound — as, for example, with the appearance of the grasses. Not only did this make possible the grazing habit of many vertebrates, but by clothing vast areas of the earth's surface it substantially reduced the sediment supply to the oceans.

11.5.1 Angiospermopsida (?Cret.–Rec.)

At present, the most abundant group of trachaeophytes, the flowering plants, vary in size from tiny ephemerals to giant trees. All, however, are characterized by having seeds which are protected by a thick coat formed from the wall of the ovule. The microsporangia (**anthers**, borne on slender filaments) often occur together with the **carpels** (female organs,

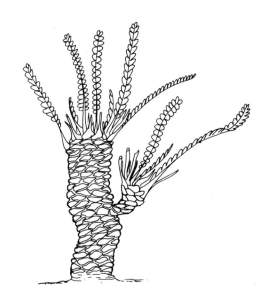

FIGURE 11.7. Reconstruction of the Jurassic cycad *Williamsonia*, found in the Rajmahal Hills of India (after Sahni, 1932).

actually captive female gametophytes) on specialized structures called flowers (see subsection 10.6.1).

The flower is an adaptation to promote and encourage **outbreeding** (i.e. it ensures that self-fertilization is rare, and that a good genetic mix is maintained within the population of each species). The seeds produced by angiosperms either possess two **seed leaves** (break open a broad bean to see these) or one. On this basis the Angiospermopsida is divided into two groups, **Dicotyledones** and **Monocotyledones**. An easy way to tell them apart is to look at the leaves. Dicot leaves have a net-like system of veins (an oak leaf, for example), while monocots (like grasses and palms) have parallel veins.

Monocots are thought to derive from the dicots, and so are often presumed to be geologically newer by quite a wide margin. This is not, apparently, the case, although it is true that the early history of the angiosperms is shrouded by a woeful lack of fossil evidence. For example, although widely believed to be a Cretaceous innovation, there are possible angiosperm fossils from the Jurassic. There is even a record of a palm-like (and therefore possibly monocot) leaf from the Triassic of Colorado (*Sanmiguelia*).

But to return to surer ground, the uncontested angiosperms of the Cretaceous were woody trees, shrubs and climbers. Soft, herbaceous plants only appeared in mid to late Tertiary time. The flowers of these tough, early angiosperms were showy, waxy and complex, made up of spirally arranged elements and probably pollinated by beetles. Fossil flowers, however, must be among the rarest of all geological specimens, and evolutionary development in flowers leans heavily upon theoretical speculations about homology for its elucidation, and does not rely upon palaeontological evidence.

As we have seen from the uses of pollen-analysis (Ch. 10, Fig. 10.7 iv), angiosperms are very useful for palaeoenvironmental work. This is because they are sensitive and because many are still alive today, so that their preferences can be noted. It took the angiosperms 20 million years — not long, geologically speaking — to rise to dominance from their relatively modest condition in the early Cretaceous. But this was merely the culmination of a process of terrestrial radiation which had begun perhaps 300 million years earlier.

12.
The Meaning of Fossils

12.1 Design, function, purpose

Fossils are 'mystery objects'. Some, of course, are more mysterious than others: but the fact is that if we want to deduce anything about life habits, then we must do so from hard parts alone against a background of the most generalized environmental information.

The complexity of the problem varies from group to group. In attempting to understand the conodont apparatus, for example (subsection 10.7.2), we are hampered by total ignorance of the soft parts. Living counterparts make our task a lot easier, and if the function which we are seeking to interpret has to do with universal physical constraints — such as water pressure — then the job is easier still. Real problems would attend any attempt to interpret adaptations associated with local environmental quirks or peculiar behaviour patterns.

But how do we set about this task in a scientific way? It is all too easy to succumb to idle fantasy and unbridled speculation in this kind of work. A rigorous framework of sound technique is what we need, but it has its roots in what may be described as a commonsensical approach.

Consider the wings of the pterodactyl. We know immediately that they were wings, because they remind us of the wings of birds and bats (Fig. 12.1 i), even though in strictly anatomical terms they are very different from one another. And if there were no bats or birds (or even flying insects) in our natural world, our familiarity with aeroplanes would afford a strong hint as to what these forelimbs did in life. We know what it takes to solve the engineering problems of flight, and our ability to interpret fossil structures is limited not by our understanding of comparative anatomy, so much as our knowledge of engineering principles and of living or mechanical machines wherein these principles are successfully employed.

A functional morphologist would therefore ask the following questions.

1. Bearing in mind the environment in which the animal lived, and taking into account any knowledge I may have about the likely biology of that organism, what were the life habits of greatest significance, and how might any of the skeletal features have aided their efficient pursuit?
2. Are there living representatives of the group with which I can compare the fossil form?
3. Are there *any* living structures which may be analogous?
4. Are there any mechanical devices which may embody similar principles?

Any functional analogy which we may draw from machines or other animals must be capable of being employed by the organism under study. That is, it must fall within the functional limitations set by materials available and the biology of the fossil group. A helicopter flies very successfully, but it cannot be used as an analogy for pterodactyl flight because (among other things) rotary bearings are unknown in living creatures (except certain flagellate bacteria).

Consideration of questions 1–4 should turn up one or a number of possible functions for the mystery structure. The next step is to design (bearing in mind the biological limitations and the need for the strictest economy) a perfect engineering solution to each projected function. Each solution is called a **paradigm** for that function. Assuming that the organism is optimally adapted (and most are), then one of these paradigms should match the actual structure precisely, and it would then be reasonable to infer that this indeed was its function in life. Complications only arise when (a) strict economy may be so vital that the best possible solution is compromised to some extent; (b) the structure has to perform two functions rather than one, leading to another form of compromise; (c) the biological limitations are unusually severe, in which case a more easily derived, inferior mechanism is developed instead of the ideal solution.

You should notice some of these complications in some of the case histories discussed below. This formal approach to the interpretation of fossils was pioneered by Professor Martin Rudwick, brachiopod palaeontologist and historian of science, during the late 1950s and 1960s. But the example which we shall consider next actually derives from another distinguished student of form, D'Arcy Wentworth Thompson. His ideas on the ammonite septum were taken up

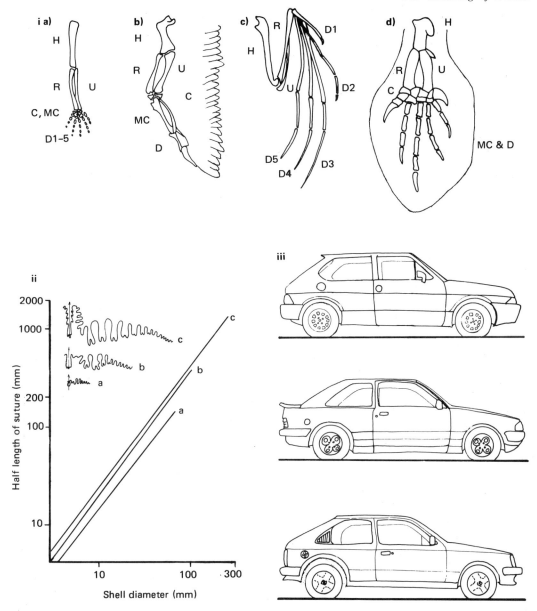

FIGURE 12.1. (i) Variations in homologous organs show how evolution has modified the same structure for different functions. (a) human arm (b) wing of bird (c) wing of bat (d) flipper of whale. Compare also the limbs of horses and litopterns (next figure). H = humerus; R = radius; U = ulna; C = carpus; MC = metacarpals; D = digits. (ii) Graph to show precise relationship existing between ammonoid shell diameter (X axis) and complexity (expressed as length) of suture. The plot is made on logarithmic scales, so the straight-line graph indicates an exponential rise in sutural length with shell diameter. The genera used are (a) *Prouddenites*, (b) *Propinacoceras* and (c) *Medlicottia* from the Late Palaeozoic. (After Newell, 1949.) (iii) Three hatchback motor cars, currently in production with Fiat, Ford and Vauxhall. Ideally, motor manufacturers would like to make their products as distinctive and individual as popular taste would allow. In recent years however, the consideration of aerodynamic function in body design has, in combination with other factors, conspired to standardize shapes about a certain 'paradigm'. This standardization is perhaps even more striking in the small hatchback range.

by a German palaeontologist, Adolf Seilacher, who used Rudwick's method in a particularly fruitful way.

12.2 The function of the ammonite septum

As ammonites grew, their shells expanded to accommodate them in a generally regular, logarithmic fashion. The soft parts moved forward in the living space, walling off the old shell volume by secreting a septum posteriorly at each removal. The chambers between septa were probably filled initially with fluid, which was drained to leave a space filled with gas at rather less than atmospheric pressure (subsection 6.17).

The septum was not flat, nor was it simply curved, but fluted marginally, joining the interior of the shell along an extremely complex suture line. The complexity of folding and the closeness of packing of the septa increased as the shell enlarged (Fig. 12.1 ii). D'Arcy Thompson was not able to explain this feature, but he suggested that the problem might be approached experimentally 'by bending a wire into the complicated form of the suture line and studying the form of the liquid film which constitutes the corresponding surface *minimae areae*' (of minimum area). In other words, make a wire loop the same shape as a suture and dip it into detergent, as though to blow a bubble. The film of liquid would be the most economical plane that could possibly join all the folds smoothly. Such a structure could be viewed as the paradigm for a septum which was nothing more than a wall, made with the least materials, across a set of fancy foundations.

Seilacher realized that the flutes of the septum extended inwards far beyond the minimum possible distance, and concluded from this that they must have performed some function, simply to justify the expenditure of extra materials and energy. Let us take some of the theories which had existed to date, and see which of them generates a paradigm most in accord with the real feature.

1. Flutes provided extra surface area, to enable the soft parts to be more firmly anchored in the shell.
2. Ammonites effected buoyancy control by compressing a body of air trapped between the body and the last-formed septum, the flutes acting as implants for the muscles required.
3. Septa acted like a submarine's bulkheads to strengthen the shell against implosion at depth.

Theory 1 would indeed encourage fluting, but the flutes would not be concentrated at the margins. Indeed, the fixation of the animal would in any case be most secure around the edges of the septum, and accordingly one would expect to see flutes placed where they would be most effective, i.e. at the centre.

Theory 2 suggests the flutes were muscle implants. Here we have many examples of muscle implants from living molluscs to look at, e.g. in bivalve shells, as well as in the *Nautilus*. No such feature is seen on the septa. In fact, the whole theory appears to be somewhat unwieldy, since every time the animal shifted forward in its shell, the entire buoyancy-regulation mechanism would be put out of action for as long as it took to make a new septum.

To strengthen a hollow body against compression, as suggested by theory 3, we might arrange a system of radiating internal struts and cross-bracings. But this solution would lie outside the limits of cephalopod biology. The problem is to design the shell so that as little of its exterior as possible goes unsupported. By folding the septum, the compressive force acting on the phragmocone is spread over a far greater length of supporting wall, i.e. the ratio of external wall to length of suture is reduced.

In the strictest model, flutes might be carried across from one side to the other, but in round-sectioned forms this would cause geometrical problems at the centre of the septum. In any case, the need for maximum permissible economy would mean that the actual structure would lie somewhere in between the 'expensive' solution and the ultimate parsimony of Thompson's bubble.

Also, as the shell grows, the shell's inherent strength-of-shape is decreased relative to the increasing stresses. So in the younger parts of the shell we should expect to see greater sutural complexity and closer spacing. Figure 12.1 ii shows a plot of shell diameter against sutural length on log/log paper. The precise interrelation between these two parameters betrays a highly organized and functionally necessary correlation.

12.3 The persistent paradigm

Since the biology of groups like the brachiopods does not alter very significantly in any sub-group, and given that the paradigm for any particular function will, for most of these subgroups, be the same, then it follows that there may often be striking similarities between quite unrelated species living in the same way. There is, if you like, only one best way of doing something, and that conclusion will be arrived at independently by many organisms. Even representatives of different phyla may come to look the same if they share the same basic system of shell construction (brachiopods and bivalves, for instance, having two hinged valves).

To see this effect in action, consider car design. Before the need for petrol economy was forced upon us by rising oil prices, car design was a matter of aesthetics rather than function, and was extremely varied as a result. But now, cars are designed for maximal aerodynamic efficiency by computers and wind tunnel experiments. Given the basic limitations of hatchback design (four wheels, front engine, tailgate, etc.), then the structural solutions to this problem are remarkably few, leading to a striking unanimity among different manufacturers.

In biology, the apparently coincidental resemblance of unrelated organisms is called **homoeomorphy**, and the process which creates them is called **convergence**. Some astounding examples are given in Fig. 12.2 ii, iii, and we shall come across some more below as we explore the mechanism of evolution by natural selection which generates homoeomorphs. But first we shall turn our attention from engineering perfection to some of nature's 'Heath-Robinson' mechanisms, which can be quite as revealing in their own way.

12.4 Perfect and not so perfect

Paradigm theory assumes that adaptations are per-

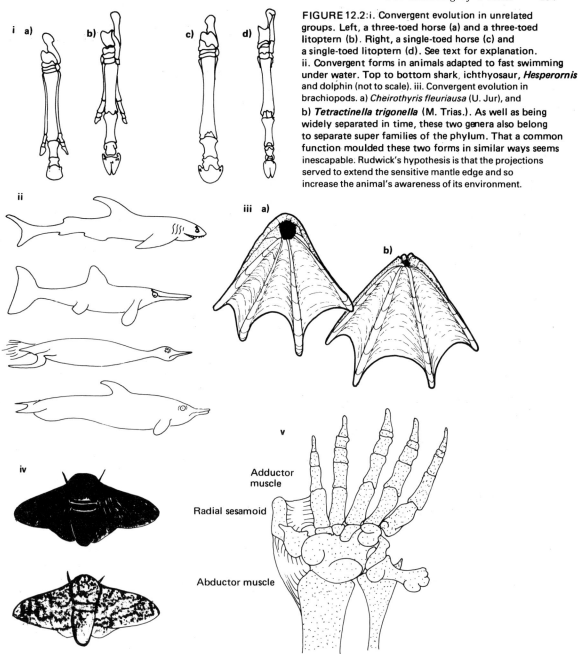

FIGURE 12.2: i. Convergent evolution in unrelated groups. Left, a three-toed horse (a) and a three-toed litoptern (b). Right, a single-toed horse (c) and a single-toed litoptern (d). See text for explanation. ii. Convergent forms in animals adapted to fast swimming under water. Top to bottom shark, ichthyosaur, *Hesperornis* and dolphin (not to scale). iii. Convergent evolution in brachiopods. a) *Cheirothyris fleuriausa* (U. Jur), and b) *Tetractinella trigonella* (M. Trias.). As well as being widely separated in time, these two genera also belong to separate super families of the phylum. That a common function moulded these two forms in similar ways seems inescapable. Rudwick's hypothesis is that the projections served to extend the sensitive mantle edge and so increase the animal's awareness of its environment.

iv Dark and normal forms of the peppered moth *Biston betularia*. The dark forms do occur naturally but are soon spotted by predators and eaten. In industrial areas however, the predominant blackening of tree trunks by polluted air placed dark forms at a selective advantage. These therefore soon exceeded their mottled kinsmen in such places as Manchester. v. The forepaw of the giant panda (from a drawing by Cramer in Gould, 1980). Owing to the uselessness of the true thumb, the bamboo-stripping requirement of the panda's new life-style has led to the creation of an apparent sixth digit from the radial sesamoid bone in the wrist. See text for implications. (Figure ii after an original in Ager, 1963)

FIGURE 12.2.

fect (or at least, nearly so) and as such it reinforces a very long-standing belief in the ultimate perfection of nature's works. But organisms are constrained by their history, and it may be that a long tradition of specialization of a particular organ can render it extremely difficult to alter — despite the fact that engineering would dictate that such an alteration would be the best course. In such cases the limitations imposed by the biology are of an extraordinarily severe kind.

The giant panda is an animal whose very existence is a marvel. Apart from an extremely restricted reproductive cycle, it is basically a carnivore which has opted for living off bamboo (which its gut is unable to digest properly). But one adaptation has been made — to strip the bamboo shoots of their leaves, the panda pulls the shoot through the gap between its paw and what looks like an opposable thumb.

But this thumb is actually a modification of a bone in the wrist (Fig. 12.2 v), because the true thumb still exists in its highly modified form, as part of the paw. It was examples such as this, highlighting the imperfect in nature, which Darwin and his supporters made much use of in defending their belief in the origin of species through evolution rather than by separate divine creation. For, as the American evolutionist Stephen Jay Gould points out, perfection of form can also be used to support the agency of divine intervention. Darwin was interested to point out structures which no sensible creator would have designed, so illustrating the importance of historical biological constraints.

If the Creator, he argued, wished to make reptiles' legs, birds' wings and whales' flippers, why did He base them all upon modifications of the pentadactyl limb (Ch. 9)? Did not the persistence of the same basic structure (Fig. 12.1 i) point to the fact that they all derived from common ancestors? God would surely not be constrained by such obsessions — a point which the skeletal morphology of angels illustrates admirably.

Angels are conventionally shown in Western art with wings *and* arms, the wings apparently attaching to the axial skeleton somewhere in the region of the thoracic vertebrae. Now if angels were earthly creatures which had evolved naturally, then their wings would have had to develop from some pre-existing structure. In all vertebrates which have developed wings, this has meant that the forelimbs have been modified to suit the function. God, however (or the imagination of man), has enabled the angels to rise above these considerations, but it remains true that for angels to have wings without divine aid, they would have to lose their arms (and forfeit the ability to play the harp).

Let us now examine the mechanism of evolution, bearing in mind what we have discussed about functional morphology and the need for adaptations to fall within the biological potential of organisms, as dictated by basic organization and historical development.

12.5 Variation and natural selection

Within animal and plant populations there exists a wide range of minor variation on either side of the norm for each species. Mature humans may range between about 2 and nearly 9 feet in height, but the vast majority cluster about 5 or 6 feet. Similar distributions exist for such parameters as head size, eye spacing, length of limbs, and so on. There is also variety in coloration of eyes, hair and skin.

The variability of natural populations reflects variations in genes, the inheritance factors which control our development. Genes are made of DNA, and occur in many bundles or chromosomes within cell nuclei. Their make-up is a kind of coded blueprint, a programme which contains all the information needed to govern every facet of bodily development from conception to death. During sexual reproduction the genetic complement of each parent is shared in the construction of the offspring, so we can have some characteristics of our mothers and some of our fathers.

But not all characteristics are inheritable. If Mr Universe were to have a son, the baby's muscles would be no better developed than those of any other, and a palaeontologist's children are born quite unburdened with any knowledge of their parent's subject! These things are called acquired characteristics, and they are never transmitted because Mr Universe's exercises and the palaeontologist's studies have no effect upon their genes.

Sexual reproduction is important in that it mixes and matches genes from all over, creating a continuous source of natural variation. It is also beneficial in that, by encouraging this mixing, faulty genes from one parent are not likely to find faulty counterparts in the other parent's genetic complement, and so are unlikely to lead to the formation of malformed offspring. It is often found that repeated intermarriage in small communities encourages this kind of mutation. The 'village idiot' in isolated rural villages, and at the other end of the social scale the various deformities which characterize many noble families (e.g. the 'Hapsburg Lip'), are good examples of this. In genetic parlance, breeding stocks are referred to as **gene pools**, to represent the total genetic complement of all breeding individuals in a species population. Small gene-pools will enter our discussion again later.

So natural populations vary widely, and most of the variants are as capable and fit as any other — their variations are neutral. In the case of severe deformities, of course, these may of themselves be directly lethal, but many may lead to death through indirect means. A gazelle with smaller or inefficient leg muscles will easily fall prey to predators, and the faulty genetic code which built the animal will be eliminated from the gene pool, and will not be handed on.

This is how, under stable conditions, the pressures of predation, space, food-supply and so on maintain the *status quo* by wiping out defective aberrants. It is referred to as **natural selection**.

12.6 Natural selection under changing conditions

When conditions alter, a new set of controls begins to act on the population, and like a slightly different sieve it begins to favour organisms which possess slightly different features. New attributes are found to be at a premium, and their possessors are favoured to survive longer. On the many hundreds of thousands of occasions when each being's fitness is put to the test, those with the newly desirable facets will tend to come off better. By their prolonged survival, their chances of passing on their characteristics are increased. So, by being more consistent in reproduction, and as a result of the elimination of other forms, the new features gradually spread among the general population, generation by generation.

The artificial selection which man employs to breed fatter cattle and faster pigeons and so on is essentially the same process, only intensified. This intensification is the result of man's efficient selection of breeding stock, which is obviously more effective than merely relying upon the 'weighted chance' which nature employs. Secondly, man's breeding stocks are small, they are undiluted by normal cattle or normal pigeons, so the variations can be made to spread and become established more quickly.

But intensive breeding is, essentially, treading a tight-rope between maximizing the degree of improvement in each generation while at the same time avoiding congenital deformities, as we have seen. And the increased incidence of deafness in some highly bred dogs shows that this is not always successful.

Of course, artificial breeding has not produced new species, all breeds of dog can (in theory, and often in practice) produce viable, but mongrel offspring. However, it is not difficult to appreciate how continuation of the process by long isolation of these breeds could lead eventually to **reproductive incompatibility**.

So, small gene-pools are good for preserving and for allowing the quick establishment of genetic changes brought about by natural selection, because the conservative inertia of a large gene-pool does not have to be overcome. The isolation of such small populations is also a prerequisite for the eventual attainment of reproductive incompatibility with the ancestral species. Such isolated, small, peripheral populations might, for example, inhabit islands or small seas or ponds and lakes. Or, like the salmon, they might be restricted to certain rivers where they return each year to spawn. Alternatively, tightly regulated social groups such as rabbit colonies may be viewed in this light.

But small gene pools also encourage mutation. As has been said, these are usually lethal and natural selection will ensure that they do not survive. But, very occasionally a mutation may be of benefit in that it could enable a form to start off on a completely new way of life — say burrowing, in a beast which had hitherto been epifaunal. Subsequent development might follow a more gradual selective pattern, but a relatively sudden key mutation could 'start the ball rolling'. The relative importance of gradual selective change and sudden mutation is discussed below in section 12.9.

12.7 Some evolutionary case-histories

12.7.1 Industrial melanism in the peppered moth

In 1850, in the English industrial heartland of Lancashire, it was noticed that the common peppered moth *Biston betularia* (Fig. 12.2 iv) occurred locally as a much darkened form which was dubbed *Biston betularia* var. *carbonaria*. And it was not long before more dark varieties of other common moths were being described, sixteen species in all, from widely separated families.

All these species shared with the peppered moth the habit of settling upon tree-trunks. Under normal conditions, mottled grey wings were a good camouflage on lichen-covered bark, but near industrial centres pollution had killed the lichen and blackened the bark. So in these areas moths with normal colouring were easily spotted by birds. Selection pressure had shifted.

Natural populations of *B. betularia* normally contain, as part of the standard intraspecific variation, some individuals with nearly black wings. Normally their life expectancy would be low, and their coloration would constitute an indirectly lethal mutation. But in Manchester the normal situation was reversed. More and more black moths survived predation, while more and more light ones died. The genetic balance of the resulting population therefore altered in favour of the black form.

12.7.2 Evolution in Cretaceous irregular echinoids

Few sedimentary environments can have been as stable as that of the Cretaceous chalks in S.E. England. Yet the form of the irregular echinoid *Micraster* changes progressively through the six zones of the Upper Chalk — a fact first noted by A. W. Rowe in the 1890s. More modern work on this same succession, and even more detailed sampling than Rowe performed, has revealed that what had long been thought of as a classic 'gradualistic' lineage does indeed have its anomalies and breaks. Several of the species comprising it now appear to have overlapped in time, and were therefore in competition with one another during such periods.

But despite the probably sudden nature of their speciation and geographic spread, these species were undoubtedly *related*, and the replacement of the earlier forms by later ones is due to their having features which allowed greater burrowing efficiency. What are these features, and how did they change with time?

Firstly, we find that through the succession, echinoid tests become wider, with the widest point and the point of maximum height migrating posteriorly. As this trend continued, the anterior ambulacral groove became deeper and more richly endowed with tubercles. The mouth shifted from a more central position on the oral surface to a more marginal, anterior one,

160   *Palaeontology — An Introduction*

while at the same time gaining a lip (the labrum) which overhangs it like a hood. Other minor changes also occurred, such as the broadening of the sub-anal fasciole and increased ornamentation in the area of the respiratory tube-feet.

The lineage of successive species is shown in Fig. 12.3. The first species, *M. leskei*, was a small, globose form, having a shallow anterior groove and a mouth which was central and lacking a labrum. These characteristics, together with the small anal fasciole and petals, suggest a form which was rather poorly adapted to burrowing (section 7.5) and to the need to breathe, feed and excrete efficiently under a cover of fine, chalky sediment.

From this species the conservative *M. corbovis* derived, retaining many of the features of *M. leskei*. Another line, however, led to *M. decipiens*, in which we see structural adaptations towards burrowing.

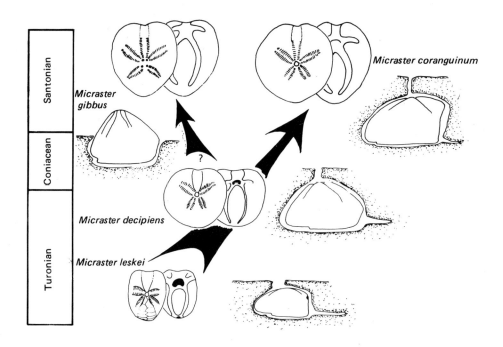

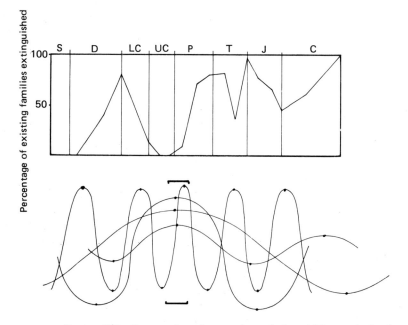

FIGURE 12.3. *Above*, modern views on the evolution of *Micraster* in Southern England. Profiles indicate depth of burial during life. (From Kermack, 1954; Stokes, 1975; Ernst, 1972; and others.) *Below*, the periods of mass extinction as expressed in the extinction of families of ammonoids: see text for details (from data in Newell, 1967). *Bottom*, natural cycles may create a concerted effect when by chance they should happen to fall 'in phase' — see bracketed area. Note close coincidence of four wave-crests. Such an event could result in severe climatic disturbance and may lead to extinction of some affected groups (after Ager, 1973).

These adaptations appear to reach their peak in *M. coranguinum*.

Somewhat abruptly, at the top of the sequence, appears another species, *M. gibbus*, a form which probably derived from *M. decipiens*. It lacks an anal fasciole, and seems to have lived semi-infaunally as a 'ploughing' echinoid. Its sudden appearance suggests a somewhat longer period of separation from the ancestral species, possibly within a peripheral population where its speciation took place.

So, within this constant environment, natural selection was operating to improve the efficiency of these echinoids as burrowers, rather than effecting a change of habit through an alteration of conditions. Close scrutiny of the fossil record reveals that speciation probably took place in peripheral isolated populations, to spread rapidly thereafter. This process produced many *relatives* (such as *senonensis* and *cortestudinarium*) but few *intermediates* (see section 12.9).

12.7.3 The evolution of horses

During the Tertiary, climates altered and forests which had clothed much of the land gave way to open grassland. It was against this background that our own species evolved (section 9.8), but it also saw the appearance of the hoofed, grazing mammals capable of galloping across the newly opened spaces at high speed. Among these were the horses.

Fossil horses are much rarer than fossil echinoids, but surprisingly good records do exist — some from the London Clay, but most from the more complete Tertiary section of North America. The ancestor of *Equus*, the modern horse, was *Hyracotherium* — a rather dog-like creature which had four toes on each front leg and three on each rear. Every one of these toes bore a small hoof, and the process of evolution towards *Equus* involved the progressive reduction of lateral digits with the emphatic enlargement of the central one.

Hyracotherium probably foraged in forest and scrub, eating leaves and fleshy shoots. Its wide feet were able to support it on marshy ground and its teeth included incisors, canines and some low-profiled cuspate cheek teeth, none of which can deal effectively with grass.

You can easily appreciate the result of millions of years of horse evolution by first placing your hand upon the table, resting firmly on the four fingertips. As horses moved on to the plains, their posture changed to an increasingly 'tip-toed' one, as an adaptation to hard ground, flat country and fast movement. Now raise your hand into the vertical position, and see how the fourth and index fingers steadily lose contact with the ground.

This loss of contact rendered the second, fourth and fifth digits useless, and they became degenerate or atrophied. In modern horses (except the rare throwback) these digits only exist as the **splint bones**, high above the hoof. The process of reduction is clearly seen in a series of intermediate species throughout the Tertiary. But did these selection pressure changes also affect other groups?

We have said that they affected the apes, and probably led to the rise of man's direct ancestors. But in similarly-organized groups such as the extinct litopterns, a similar process of digit-reduction took place. It culminated in *Thoatherium* (Fig. 12.2) which, despite the striking homoeomorphy with *Equus*, is quite unrelated to horses.

Living chiefly in South America, *Thoatherium* occupied (somewhat earlier) the same ecological niche as *Equus* came to do elsewhere. But the eventual extinction of these otherwise well-adapted plains-dwellers may have had more to do with their teeth than their toes.

In horses, a parallel dental evolution changed the unspecialized teeth of *Hyracotherium* into the high-crowned molars needed to deal properly with large quantities of tough, abrasive grass. Litopterns never made this modification, and it is thought that this may have led to their ultimate failure.

So far, we have approached evolution through the practice of functional interpretation, to highlight the fact that evolution proceeds through the testing, by natural selection, of a wide range of structures which are vital to the survival of any organism. In other words, the principle of evolution by natural selection entitles us to regard every structure as adaptive, as functional.

We shall now see that there are logical reasons for this standpoint too, as we consider some evolutionary models which have placed at least some faith in non-adaptational or **orthogenetic** processes.

12.8 Evolutionary trends and orthogenetic theories

The possibility that rare chance mutations may meet with success has been mentioned, but evolution is never 'forced'. Natural selection will always ensure that all structural modifications are tested by the same standard. This selection is a product of all the myriad processes going on inside the ecosystem, comprising space, predation, competition and availability of food. No evolutionary trend, unless it operates solely upon neutral characters of no survival importance, can be seen as anything other than an opportunistic response to a change in local conditions.

In the past, however, many apparently 'bizarre' structures were interpreted in the light of orthogenetic theories which had in common the notion that the trends seen betrayed the working out of inherent tendencies within the genome (each individual's genetic complement). Such theories talked of internal 'drives', or 'evolutionary momentum' which could carry organisms past the point of optimum adaptation into regions of positively disadvantageous 'overdevelopment'. Some theories also pointed to the supposed prevalence of bizarre forms at the ends of lineages just prior to their extinctions, and suggested that they had become racially 'senile' or 'decadent'.

This last notion, deriving less from the evidence than from analogies with individual old age and thoughts of the last days of Rome, was much favoured as an explanation of the heteromorph ammonites (section 6.17). They were described as '. . . biologically absurd . . . ridiculous for the basic concept of ammonite form' (Daqué, 1935). So the strangely uncoiled forms were non-functional signs of a kind of formal incontinence.

But this explanation falls down on close examination. To begin with, only a few forms ever uncoiled, and the existence of one or two such genera could hardly account for, as an example, the extinction of all ceratitids at the end of the Triassic. Moreover, many Cretaceous uncoiled forms gave rise to properly coiled descendant species, showing that their inherent ability to produce normal shells was not lost from their genes — merely suppressed. The reason for this is that the heteromorph forms had a distinct functional advantage in certain modes of life (section 6.17) and indeed many of these genera were extremely successful — hardly what we should expect of gerontic incompetents tottering towards extinction.

Orthogenetic theories not only run counter to the principle of evolution by natural selection, they are also logically unsupportable. Rudwick has pointed out that while it is possible to support or disprove the assertion that any fossil structure was functional, it is always impossible to validate the opposite standpoint. There always remains the possibility that the supposedly useless structure possessed some purpose which we have either not the perspicacity or the necessary information to see.

So functional morphology lies at the very heart of evolutionary palaeontology. The phenomenon of homoeomorphy attests the validity of the functional approach, as it also provides awesome proof of the efficacy of natural selection. And the more we examine the subject, the more it becomes clear that the principle of evolution by natural selection embodies an inevitable life-process, making it not a force operating within nature, but one of its fundamental definitions.

12.9 Small steps and giant leaps

That the fossil record provides us with few gradual evolving lineages has already been suggested. While gradualistic changes are, like those of the moth *Biston betularia*, demonstrable in the present and known from rare fossil examples, we usually witness sudden appearances of new species, followed by long static periods and ultimate extinction.

Formerly it was supposed that this fact reflected the incompleteness of the fossil record, but the belief now is that it represents something very important about the evolutionary process. Gradual phyletic transformation of one species into another is probably one minor mechanism, while most species evolve quickly in the peripheral, isolated communities which we have talked about before. Invasion from these 'crucibles' is a geologically instantaneous process, leading to species with many likely relatives, but no material 'missing links' (subsection 12.7.2). The way in which periods of stasis are broken by abrupt incursions of this kind has led to this style of evolution being dubbed **punctuated equilibrium** by its chief advocates, Niles Eldredge and Stephen Jay Gould. It does not replace the alternative **phyletic gradualism**, but stands alongside it, explaining the observed features of the fossil record much more successfully while itself receiving support from the study of genetics.

But so far what we have been talking about has been micro-evolution, at the fundamental level of the species. But major shifts in the organization of life may not necessarily be this micro-evolution 'writ large' — nor is it vital to the principle of Darwinian evolution that they should be.

Part of the problem with large-scale developments like the wing of a bird or the reptilian sealed egg (Ch. 9) is that their evolution presumably involved stages which were only partly formed, and therefore useless. What selective advantage to a bird-like reptile would a non-functional wing be? The problem is that for natural selection to perform the transformation, there must be some benefit conferred all along the way. A half-bird cannot beg nature for just a little more time to allow it to develop fully, as though it were some struggling research-and-development project asking its bank manager to be patient. In the absence of orthogenesis (which would say that the trend could be directed by some internal drive) we must look elsewhere for an explanation.

It is very likely in cases such as these that the wing that now enables birds to fly, and the egg which enables reptiles to live on land originally developed for entirely different reasons. *Archaeopteryx* could have used its wings to entrap prey — as many modern birds still do. Only later was the functional potential of a wing exploited — first for gliding, and then, with the development of stronger muscles and lighter bones, for powered flight (section 9.6).

Similarly, as discussed in Chapter 9, the sealed egg is of positive adaptive value to water-dwelling reptiles like the turtle and the crocodile. It could, therefore, be evolved for purely adaptive reasons before its potential for freeing reptiles from the water was exploited by the first terrestrial forms. This so-called **pre-adaptation** (section 9.5) certainly appears to have been an important force in vertebrate evolution. It is also an important concept in the explanation of how Darwinian processes can give rise to complex, perfect structures through all their incipient stages.

12.10 Extinction

From the origin of species we move at last to their ultimate fate. Of course, if one species evolves completely into another such that a palaeontologist deems them to be different enough to be given separate names, then the older species becomes extinct. But this is **phyletic extinction**. What we wish to consider here is nothing less than eradication — the total removal of one species' contribution to life on earth. For this is no rare event. On average, species tend to last for one or two million years and then vanish, with few exceptions. Why is this?

The usual run-of-the-mill extinction of the odd species here and there, a process which is going on all the time, is generally the result of either drastic environmental changes to which the organism cannot adapt, or to evolutionary developments elsewhere. Such a lethal development might, for example, be the appearance of a super-efficient competitor in the same ecological niche, or the extinction of a favoured food species, or of a species with which there was a vital ecological link. Alternatively, there might evolve a particularly rapacious predator. The appearance of *Homo sapiens* has rung the death knell of many hundreds of species other than the unfortunate Dodo.

But in contrast with this type of extinction, there is the **mass extinction**. If you have been reading through the 'geological history' sections of the preceding chapters, you will have noticed that very many different lineages end or begin at a few very significant moments in geological time. Some of these biological crises were more disastrous than others, but the history of life reveals many such discontinuities.

Breaks of this kind are seen at the end of the Cambrian, Ordovician and Devonian periods. A very significant one delineates the close of the Permian, and there were more at the ends of the Triassic and Cretaceous. The fact that they seem to mark the boundaries of geological systems is no coincidence, but what is remarkable is the scale of many of these events. Let us look at the end-Permian extinction, since it was, perhaps, the greatest, and there is at least one plausible explanation for it. For the others, explanations are perhaps as numerous as the species which succumbed to them, ranging from oil pollution to meteoric impacts and the existence of mysterious companion stars to the sun. Alas, the relative merits of these fascinating speculations lie outside the scope of this book.

Some 225 million years ago, 52% of all families of marine organisms vanished. On the species level, this could mean that over 90% were decimated (families are by no means all the same size), so it is not surprising that the survivors dictated the entire character of Mesozoic life. Against this sort of pressure, there can have been little that any organism could 'do', in an adaptive sense, to ensure survival. It would appear that in these circumstances the strong went down with the weak and continuation was a matter of good fortune rather than good biology. Why did it happen?

Today the continents are riding fairly high upon their lithospheric plates and are mostly racing away from each other. All around their edges are narrow, but linearly extensive shelf seas where sea-level overlaps their margins. It is in these areas that most marine life is concentrated — most, certainly, in terms of what gets preserved in the fossil record.

At the close of the Permian, however, all the continents collided with one another, fusing into a supercontinent called **Pangaea**. The results upon climate, oceanic circulation and the world ecosystem as a whole must have been immense — but nowhere were they more severely felt than upon the continental shelf. Firstly, by colliding and fusing, the actual length of continental margin was drastically reduced. On top of this, the cessation of ocean floor spreading meant that the spreading centres collapsed. These vast submarine mountain ranges, like the present-day Mid-Atlantic Ridge, sank down, so increasing the volume of the ocean basins that they caused a massive regression which narrowed the shelf seas even further.

This meant that there was very little space to support the marine biota in these areas, and it is one of ecology's cardinal rules that the more restricted the space available, then the lower the diversity of species which it can support. The conclusion must be that the upshot of this continental crash was nearly the total destruction of marine life.

But even in this case it is likely that area was not the only factor. And we cannot correlate all mass extinctions with the formation of supercontinents. The rest must await developments, just as the Permian extinction had to wait for plate tectonic theory to receive its first tentative acceptance. All that can be said in the meantime is that one cause alone is almost never the reason for anything in geology, and that we must look to the interaction of many causes (terrestrial and extraterrestrial) whose rough cyclicities might come very infrequently into phase (Fig. 12.3) and produce one concerted assault upon the whole edifice of life.

Suggested Further Reading

In Greater Detail

Ager, D. V. (1963) *Principles of Palaeoecology*. New York: McGraw-Hill. The first textbook in English on the subject. Useful and readable basic account.

Andrews, H. N. (1961) *Studies in Palaeobotany*. New York: J. Wiley & Sons Inc. An excellent and well-illustrated guide, with an introduction to palynology by Charles J. Felix included.

Brasier, M. D. (1979) *Microfossils*. London: George Allen & Unwin. Invaluable and long-awaited textbook on all major microfossil groups, with accounts of preparatory techniques.

Clarkson, E. N. K. (1979) *Invertebrate Palaeontology and Evolution*. London: George Allen & Unwin. Quite the best advanced textbook of palaeontology available. The origin of life, evolution and the major invertebrate groups all described.

Kennett, P. and Ross, C. A. (1983) *Palaeoecology*. York: Longman Group Ltd. Admirable large format case-history treatment, aimed at 'A' level students and reasonably priced.

McKerrow, W. S. (1978) *The Ecology of Fossils*. Duckworth. Survey of fossil communities through the British stratigraphical record.

Moore, R. C. (and Teichert, C.) (eds.) (1953) *Treatise on Invertebrate Palaeontology*. Geol. Soc. America and Univ. Kansas Press. Standard reference on all invertebrate fossils. Each phylum is treated separately, each major fossil group being awarded an individually lettered volume. Useful source of biological background-information also.

Raup, D. M. and Stanley, S. M. (1978) *Principles of Paleontology*, 2nd edn. San Francisco: Freeman. Exciting and still novel textbook treating the concepts and methods of the science rather than the morphology and stratigraphy of fossil groups. Rapidly becoming a modern classic.

Romer, A. S. (1959) *The Vertebrate Story*, 4th edn. Chicago: University of Chicago Press. Perennially popular and readable, this easy-going account of the evolution of vertebrates serves as the best possible introduction to a fascinating topic.

Rudwick, M. J. S. (1970) *Living and Fossil Brachiopods*. London: Hutchinson & Co. Summation of his brilliant researches into the Brachiopoda, this specialized treatment is at once authoritative and readable. Revolutionary in its approach, this book showed what could be done with a single-group, palaeobiological text.

Smith, A. (1984) *Echinoid Palaeobiology*. London: George Allen & Unwin. A worthy successor to the above, concerned with the true understanding of fossil form and distribution.

Biological Background

Barnes, R. D. (1974) *Invertebrate Zoology*, 3rd edn. Philadelphia: W. B. Saunders & Co. Invaluable textbook on major, minor and even some fossil invertebrate phyla. Authoritative, readable and beautifully illustrated by photographs and drawings.

Buchsbaum, R. (1951) *Animals Without Backbones* (2 vols.). Pelican. A much-reprinted introduction to the invertebrates, concentrating on their construction.

General Reading

British Palaeozoic Fossils, *British Mesozoic Fossils* and *British Caenozoic Fossils*. British Museum of Natural History (5th edn., 1975). Survey of the more common British fossils, exquisitely illustrated. Invaluable and essential handbooks.

Thompson, D'Arcy Wentworth (1917) *On Growth and Form*. (abridged edn. J. T. Bonner (ed.) 1961; original in two volumes). Cambridge: Cambridge University Press. Idiosyncratic and strangely inconclusive work of genius, treating the physical laws which govern animal form and the mathematical expression of the products of growth. Written with all the dignity of the best Victorian prose style.

Gould, S. J. (1980) *Ever Since Darwin*. Pelican. Originally published in America, this is the first of Gould's collections of essays to be reissued in Britain. Thought-provoking, easy-going meditations on evolutionary biology, palaeontology, science and society and science history. Cannot be too highly recommended.

Gould, S. J. (1980) *The Panda's Thumb*. New York: W. W. Norton & Co. Inc. More of Gould's 'reflections in natural history'. Witty and compelling reading.

Gould, S. J. (1983) *Hen's Teeth and Horse's Toes*. New York: W. W. Norton & Co. Inc. Latest of Gould's collected essays.

Author Index

Adanson, M. 10
Ager, D. V. 29, 157, 160
Ananiev, A. R. 147
Andrews, H. N. 149

Bambach, R. K. 53
Brasier, M. D. 134
Bütschli, C. 134
Bystrow, 123

Calman, W. T. 141
Carruthers, R. C. 113
Claus, 141
Cobban, W. A. 77, 82
Cocks, L. R. M. 53
Cope, J. C. W. 82

Daque, E. 162
Darwin, C. 118
Dean, B. 124

Eldredge, N. 162
Ernst, G. 160

Favre, 137
Fenton, C. L. 77
Fenton, M. A. 77
Fischer, A. G. 101, 117
Flower, R. H. 81
Fox, S. 12

Gale, A. S. 103
Graham, A. 68
Grand D'Eury, C. 152
Gould, S. J. 157, 158, 162
Grove, A. J. 68

Hantken, 137
Heilmann, G. 128
Høeg, O. A. 145
Hyman, L. H. 111

Jarvik, 126

Kennedy, W. J. 77, 82
Kermack, K. 160
Kesling, R. V. 141
Kidston, R. 144, 145
Kier, P. M. 104
Krausel, R. 145

Lalicker, C. G. 101, 117
Lang, W. H. 144, 145
Linnaeus, C. 9, 10, 84

Makowski, H. 82
Manton, S. M. 16
Mark, 123
Markhoven, 141
Michael, 137
Miller, J. J. 12
Moore, R. C. 22, 74, 85, 101, 117
Mori, K. 119
Murchison, Sir R. I. 9

Newell, G. E. 90, 91
Newell, N. D. 155, 160
Nichols, D. 103
Nield, E. W. 120

Oparin, A. 13

Palframan, D. F. B. 77
Peel, J. S. 73

Raup, D. M. 67
Romer, A. S. 127
Rowe, A. W. 159
Rudwick, M. J. S. 47–8, 51, 54, 154–5, 157

Sahni, B. 153
Seilacher, A. 155–6
Smith, A. B. 102–3
Stensiö, E. 123
Stenzel, H. B. 74
Stokes, R. B. 160

Teichert, C. 120
Thompson, D'A. W. 131, 154–6
Trueman, Sir A. 81
Trueman, E. R. 90
Turner, C. 139

Vaughan, T. W. 118
von Linné, C., see 'Linnaeus, C.'

Walton, J. 149
Weyland, H. 145
Williams, A. 48

Zeigler, A. M. 53

Systematic Index

Acanthograptus 37
Acaste 20, 24
Acervularia 115
Acidaspis 24
Acrotretida 55, 59
Actinocrinus 109
Actinomma 134
Actinoptychus 134
Agnatha 121
Agnostida 24, 26
Agnostina 24
Agnostus 24
Alethopteris 149, 151
Alveolina 137
Ammonoidea 74–5, 79, 83, 86–9
Ammonitida 77, 79, 83, 85, 88–9
Amphibia 121
Anarcestida 85–6
Ananaspis 25
Angiospermopsida 152
Animalia 9, 130, 140
Anisograptidae 38
Annularia 147–8
Anthozoa 110
Anthraconaia 98
Anthraconauta 98
Anthracosia 98
Anthropoidea 9
Apiocrinites 109
Arachnida 9, 16
Arachnophyllum 8
Araucaria 150
Araucarioxylon 150
Arca 95
Arcestes 87
Archaeogastropoda 67
Archaeopteryx 127–8, 163
Arcoteuthis 84
Arthropoda 16, 64, 130, 140
Articulata 44, 56, 59
Asaphina 24
Asaphus 24
Asteroidea 99
Asterotheca 151
Asteroxylon 144–5
Athyridina 59
Atrypa 56, 59, 62
Atrypella 59
Atrypidina 45, 56, 59, 62
Atrypina 59
Aulophyllum 114
Aves 121

Baculites 77–8
Baragwanathia 145
Bathynotus 29
Bathyuriscus 24
Bathyurus 24
Belemnella 84

Belemnitella 84
Belemnitida 79
Bellerophontacea 71–2
Bennettitales 151–3
Binatisphinctes 89
Biston 157, 159, 162
Bivalvia 64, 84
Blastoidea 99
Bonnia 29
Buccinum 68
Bumastus 24, 28

Caenogastropoda 67
Calamites 147–8
Calceola 115, 117
Callavia 29
Callograptus 37
Calymene 24, 26
Calymeniina 24
Camarotoechia 59
Canis 8
Carbonicola 98
Cardioceras 89
'Cardium', see 'Cerastoderma'
Carneithyris 59
Cephalaspis 122–3
Cephalopoda 64, 74
Cerastoderma 92–4
Ceratias 82
Ceratites 87
Ceratidia 77, 83, 85, 87
Ceratocephala 24
Charnia 14
Cheirothyris 157
Cheirurina 24
Cheirurus 24
Chondrichthyes 121, 124
Chonetes 49, 53–4
Chonetidina 59
Chordata 9, 99, 121
Chrysophyta 132
Cidaris 105
Cladoselache 123–4
Clavilithes 71
Climacograptus 35, 40
Clisiophyllum 114
Clorinda 53
Clymeniida 85–6
Clypeus 106
Cnidaria 110
Coleoidea 74–5, 79, 84
Conchidium 45, 50
Conodontophorida 130
Conus 72
Cooksonia 144–5
Cordaitales 148, 152
Corynexochida 24–5
Cosmoceras 82
Crania 50, 56, 59

166

Craspedobolbina 142
Cremiceras 77
Crepidula 72
Crinoidea 99, 106
Crinozoa 99, 106
Crustacea 16, 140
Cryptograptidae 35
Cryptograptus 35, 43
Cryptolithus 24
Cycadales 151–2
Cyclococcolithina 134
Cyclomedusa 14
Cyclopteris 151
Cyclopyge 23
Cymaclymenia 86
Cyrena 93
Cyrtia 59
Cyrtograptidae 35
Cyrtograptus 35
Cystoidea 99

Dactylioceras 81, 88
Dalmanites 23–4
Deiphon 24
Dendroidea 30
Dibunophyllum 114, 118
Dicellograptus 39
Dichograptidae 35, 38
Dickinsonia 14
Dicotyledones 153
Dicranograptus 35, 39, 40
Dictyonema 31–2, 36–7
Didymoceras 82
Didymograptina 35
Didymograptus 33, 35, 38
Dielasma 59
Digonella 59
Dimorphograptidae 35, 40
Dimorphograptus 35, 40
Dinichthys 125
Dinoflagellata 130
'Dinosauria' 127
Diodora 68
Diplograptidae 35, 40
Diplograptina 35
Diplograptus 33, 39
Diploria 115, 117
Discinisca 52
Drepanophycus 145

Echinocardium 102–3
Echinocorys 106
Echinodermata 99
Echinoidea 99
Echinus 100–1
Encrinurus 108
Enigmophyton 144–5
Ensis 92–3
Eocoelia 53
Eodiscina 24
Eodiscus 24
Equisetum 147–8
Equus 161
Ernestiodendron 150, 152
Eryops 126

Favosites 113, 116, 120
Filicopsida 148
Foraminiferida 130, 132
Fordilla 98
Fungi 130

Gallus 128

Gastrioceras 83
Gastropoda 64, 67
Gervillia 96
Gigantoproductus 59
Ginkgo 150–2
Globigerina 134
Glossograptidae 35, 40
Glossograptina 35
Glossograptus 35, 40
Glycimeris 95
Glyptograptus 40
Gnathostomata 121
Goniatites 77
Goniatitida 79, 83, 85–6
Goniorhynchia 63, 69
Gonioteuthis 84
Graptolithina 30
Graptoloidea 30, 35
Gryphaea 121
Gypidula 59

Hagenowia 103
Halysites 113, 116, 120
Hamites 88
Harpes 24
Harpina 24
Hedbergella 137
Heliolites 116, 120
Helix 66
Hemichordata 30, 36
Hesperornis 157
Heterorthis 59
Hibolites 84
Holmia 29
Holoptychius 124
Holothuroidea 99
Hominidae 9
Homo 8, 163
Hyracotherium 161
'*Hypsiprimnopsis*' 127

Ichthyosaurus 126–7
Illaenina 24
Illaenus 24
Illinella 137
Inarticulata 44, 55–6, 59
Insecta 9, 16
Isastrea 117

Ketophyllum 115
Kjerulfia 29
Kosmoceras 89

Laevaptychus 82
Lagenostoma 148–9
Lamellibranchia 84
Lampetra 122–3
Lasiograptidae 35, 37
Lasiograptus 35
Latimeria 125
Latirus 70
Lebachia 150, 152
Leonaspis 24
Lepidodendron 145–6
Lepidophylloides 146
Lepidosiren 124
Lepidostrobus 145–6
Leptaena 49, 58–9
Leptograptus 35, 38
Lichas 26
Lichida 24, 26
Liljevallia 51
Lingula 45, 47, 49–50, 52–5, 59

168 Systematic Index

Lingulella 45, 59
Lingulida 59
Lingulopsis 59
Lithophaga 96
Lithostrotion 114–15
Lobograptus 41
Lobothyris 59
Loganopeltis 24
Lonsdaleia 118
Lucina 93
Lycopsida 114
Lyginopteris 148–9
Lytoceratida 83, 85, 88

Maclurites 72
Maenioceras 86
Mammalia 9, 121
Marsupites 109
Mastigobolbina 142
Medlicottia 155
Medusinites 14
Melosira 134
Mercenaria 92
Meristina 59
Merostomata 16
Metazoa 9
Micraster 102, 104, 106, 159–61
Miocidaris 106
Modiolus 92, 96
Mollusca 64
Monera 130–1
Monocotyledones 153
Monograptidae 35, 38, 41–3
Monograptina 35
Monograptus 33, 41–2
Monotremata 127
Montlivaltia 117
Mourlonia 71
Mucrospirifer 62
Mya 93, 94, 97
Myriapoda 16
Mytilus 92, 94, 97

Nautilida 83
Nautiloidea 74–5, 83, 85
Nautilus 74–5, 80, 83, 156
Nemagraptidae 35, 38–9
Nemagraptus 39
Neotrigonia 93
Neuropteris 151
Nevadia 29
Nucula 90, 93, 98
Nummulites 73

Odontopleurida 24, 27
Ogygiocarella 26
Ogygiocaris 24
Olenellina 24
Olenellus 24, 28–9
Olenoides 24–5
Olenus 24
Onnia 24, 27
Onychocrinus 108
Ophiuroidea 99
Opisthobranchiata 67, 69, 72–3
Orbiculoidea 47, 50, 55, 59
Ornithella 61
Ornithorhynchus 127
Orthida 45, 56, 59–60
Orthograptus 35, 40
Orthonbyoceras 85
Osteichthyes 121, 123
Ostracoda 130, 140
Ostrea 92–3

Paedumias 28
Palaeosmilia 118
Panope 92
Paradoxides 24
Parawocklumeria 86
Pecten 87, 92–3, 97
Pectinatites 82
Pentagonia 59
Pentamerida 56, 59, 61
Pentamerus 45, 53, 59
Periarchus 104
Peronopsis 25
Phacopida 23–7
Phacopina 24
Phacops 24
Phillipsastrea 115
Phillipsia 24
Pholadomya 95
Pholas 91, 92, 95, 97
Phylloceratida 83, 85
Phylloceras 77
Phyllograptus 33, 35, 38
Phymosoma 105
Pinna 89, 92
Pinus 139, 150
Placenticeras 77
Placodermi 121, 131
Planta 130, 136
Pleurodictyum 117
Pleuromya 95
Pleurotomaria 73
Plectronoceras 83
Porifera 118
Primates 9
Priapulida 9
Prionocyclus 89
Productidina 45, 59, 61
Productus 57, 59
Proetida 25
Proetus 24
Prolecanitida 85, 87
Propinacoceras 155
Prorichthofenia 54
Prosobranchiata 67
Protista 130, 132
Protobranchia 84
Protochonetes 58–9
Protochordata 30
Protohyenia 148
Protolepidodendron 145
Protopterus 124
Protospongia 8
Protypus 29
Prouddenites 155
Psammolepis 122–3
Pteria 92
Pteridospermales 148
Pteropoda 72
Ptychocephalus 23, 29
Ptychoparia 24
Ptychopariida 24, 27–8
Ptychopariina 24
Pugnax 49, 59
Pulmonata 67
Punctospirifer 60
Pygaster 105
Pyrrhophyta 132

Radiolaria 130, 132
Rafinesquina 59
Ramapithecus 128
Rangea 13
Rayonnoceras 85
Redlichia 24
Redlichiida 24, 28
Redlichiina 24

Systematic Index

Reptilia 121
Retiolites 35
Retiolitidae 35, 37
Rhabdopleura 35–6
Rhacopteris 149
Rhynchonellida 45, 56, 59, 63
Rhynia 144–5
Richthofenia 59
Richthofeniida 61
Rugosa 110, 114, 117
Rugosochonetes 59

Saetograptus 41, 42
Salopina 59
Sarcodina 132
Sarcopterygii 121, 125
Sanmiguelia 153
Sansabella 142
Scaphites 88
Schizophoria 56, 59
Schloenbachia 88
Scleractinia 110, 114, 116, 120
Sclerospongia 118
Scottognathus 137
Scutellum 24
Selenopeltis 27
'*Seminula*' 118
Septibranchia 84
Sequoia 150
Shumardia 29
Sigillaria 146
Siphonophyllia 118
Skogsbergia 141
Solemya 98
Solen 92–3
Sowerbyella 59
Sphenophyllum 147–8
Sphenopsida 147
Spirifer 52, 59–60
Spiriferida 45, 56, 59–60, 62
Spiriferina 59
Spondylus 94
Spriggina 14
Stenocisma 8
Stigmaria 145–6
Stenopronorites 86
Strenuella 29
Stricklandia 53
Stromatoporoidea 118
Strophomena 59

Strophomenida 45, 56–9, 61
Strophomenidina 59
Sulculus 71
Synbathocrinus 108
Syringopora 116
Syringospira 54

Tabulata 110
Telangium 149
Terataspis 24
Terebratellidina 59
Terebratula 54, 59, 66
Terebratulida 45, 56, 59–61
Terebratulidina 59
Terebratulina 59
Teredo 92, 95–6
Tetractinella 157
Tetragraptus 33, 35, 38
Tetrarhynchia 59
Thamnasteria 117
Thecosmilia 117
Theraspida 127
Thoatherium 157, 161
Tornoceras 86
Tremanotus 71
Triarthrus 22, 24
Tribrachidium 14
Tridacna 64, 98, 132
Trigonia 93
Trilobita 16
Trimerus 24
Trinucleina 24
Trinucleus 24
Turrilites 88
Turritella 72

Vertebrata 9, 121

Waagenoconcha 54
Williamsonia 151, 153

Zaphrentis 118
Zigzagiceras 5
Zoantharia 110
Zosterophyllum 144
Zygobolbina 142

General Index

A priori 55
Abapical 70
Abductor muscle 157
Abereiddy, Wales 38
Abergwesyn, Wales 42
Aboral 99, 100
Absolute dating 5
Acado-Baltic Province 29
Accessory fossils 6
Accessory shell 96
Accessory tube-feet 105
Acetic acid 142
Acid-insoluble residue 140
Adapical 70
Adductor muscle 46–8, 50–1, 56, 84, 90–1, 102, 140–1, 157
Adjustor muscles 46, 50, 53
Adradial suture 102, 105
Aerodynamics 155
Aeroengineering 128
Aerofoils 122
Aeroplane 122
'Age of fishes' 122
Agglutinated foraminifera 135
Ahermatypic 116
A.I.R., see 'acid-insoluble residue'
Air-sacs 140, 150
Airship 122
Alabama, USA 77
Alar fossula 113
Alar septum 113
Alaska, USA 3
Alden, New York, USA 86
Alete spores 140
Algal mats 131
Allantois 126, 129
Alveolus 78–9
Amber 3
Ambitus 100–1
Ambulacrum (-cra) 100–1, 104–7
American slipper-limpet 72
Amino-acids 12, 129
Ammonite zones 5, 38
Ammonitic suture 77, 88
Amnion 126
Amniotic egg 121, 126–7
Amphibian 125–6
Amphidetic 94
Ampulla 101, 103
Anal fin 123
Anal plate 107
Anal pore 86
Anal tube 107, 109
Anisomyarian 90, 92, 94, 97
Angels 158
Angiosperms 138
Angler-fish 82
Annual rings 146
Antapical 132
Anthers 152
Anthropoid apes 129

Antennae 18, 141
Anterior 16
Apatite 142
Apex 47, 70
Apical angle 70
Apical disc 100, 104–6
Aptychus (-chi) 77, 82
Aragonite 2, 4, 69, 71–2, 77, 89, 110
Archaeopyle 132
Ardtun Leaf Beds 150
Areole 102, 105
Arisaig, Nova Scotia 73
Arisaig Group 73
Aristotle's Lantern 101–2
Arms 107
Arthrodire 123
Articulating half-ring 20
Articulating half-segment 21
Artificial selection 159
Asexual division 110, 131–2
Astorhizae 118–19
Astraeoid 112–13
Atherfield, Isle of Wight 96
Atoll 118
Aulton, Somerset 117
Auricle 94
Auriform 71
Australia 86, 93, 117
Autotheca 31, 111
Axial canal 107
Axial complex, see 'axial structure'
Axial node 20–1
Axial ring 20
Axial skeleton 121
Axial spine 21
Axial structure 112–14
Axis 16

Bacteria 12–14, 130–1
Ballast tanks 75, 179
Baltic Sea 3
Bamboo 157
Barmouth, Wales 94
Barmouth Viaduct 92
Barnacles 53
Barrier reefs 118
Bars 142
Barton Beds 73
Basal disc 31
Basal plate 110–11
Basals 107
Basingstoke, Hants 105
Bathyurid Fauna 29
Bats 129, 155
Bavaria 82, 128
Beak 46–7, 56–7, 62–3, 91
Beetles 153
'Belemnite Chalk' 84
Belemnites 74

Bellerophontiform 71
Benthic assemblage 54
Benthic cyst 132–3
Benthos 5
Berlin, Germany 41
Biconical 70
Bifid tooth 136
Bifurcate 76, 88–9
Big Horn Mt., USA 3
Binomen 8
Bioerosion 102
Biogenesis 12
Biospecies 9
Biostratigraphy 5
Bipinnate 107–8
Biramous 19, 22–3
Birds 121, 127
Biserial 33, 136
Bisulcate 76, 88–9
Bitheca 31, 111
Blades 142
Blood group 129
Blue-green algae 130–1
Body chamber 75–6, 78, 88
Body fossil 1
Bohemia 43
Bone 2, 121, 123–4
Bony fishes 121, 123
Border 21
Boss 102, 105
Boueti Beds 63
Brachial facet 107, 109
Brachial plates 107–9
Brachial valve 44, 46–7
Brachials, see 'brachial plates'
Brachidium (-ia) 44, 48–50
Brachium (-ia) 107–9
Bradford-on-Avon, Wilts 109
Brain-coral 117
Breast-bone 128
Breeding, see 'artificial selection'
Bristol, Avon 127
Brood-pouches 141
Brush tubules 48
Buoyancy 37, 79, 80–1, 122, 124–5, 135
Buoyancy, centre of, 80
Burgess Shale 20
Burlington, Iowa, USA 108
Byssal gland 98
Byssal mineralization 98
Byssal notch 94, 97
Byssus 88, 96

Cadicone 76
Caerphilly, Wales 152
Calceoloid 112, 115
Calcite 2, 4, 77, 89, 110
Calcium Carbonate Compensation Depth 132
Caledonian Orogeny 28
Calice, see 'calyx'
Callus 70
Calymma 135
Calyx 106–7, 109–10, 112, 115–17
Camera (-ae) 74–6, 78
Cameral deposits 76, 78–9, 80–1, 85
Cameral fluid 80
Camouflage 80
Canary mule 8
Carapace 2, 141
Carboniferous Limestone 57, 83, 117–18, 127
Cardinal extremity 46
Cardinal fossula 113–14
Cardinal process 50–1, 57–8
Cardinal septum 113–14
Cardinal teeth 91, 93
Carmarthen, Wales 14

General Index 171

Carpals 112
Carpel 152
Carpus 155
Carter Co, Oklahoma, USA 40
Cartilage 2, 121, 123–4
Cast 4
Catch adductor 52
Catch apparatus 102
Catherine Wheel 66, 82
Caudal furca 141
Cavate 132, 140
CCCD, see 'Calcium Carbonate Compensation Depth'
Cementation 53–4
Central nodule 134
Centrale 122
Centric 132, 134
Cephalic fringe 19
Cephalic spine 19
Cephalon 17, 65
Ceratohyal 124
Ceratoid 112
Cerebral ganglion, see 'Ganglia'
Cerioid 112, 115, 117
Chain-coral 116
Chalk 106, 134, 159
Chamber, see 'camera'
Charnwood Forest, Leics 14
Cheltenham, Gloucs 95
Chimney-brush 102
Chitin 2, 20, 44, 50, 140
Chitinophosphatic 52
Chlorophyll 131
Chondrophore 94, 97
Chorate 132
Chorion 126
Chromosomes 135
Chron 5, 38–9
Chronospecies 9
Cilia 84
Cincinnati, Ohio, USA 85
Cingulum 132
Circum-oral canal 101
Cirrus (-i) 107
Cladium (-ia) 35, 43, 87
Clam 92
Clarita, Oklahoma, USA 87
Class 9
Classification 8
Clitheroe, Lancs 108
Club-moss 144
Clypeastroid 104
Coal 2
Coal Measures 83, 98, 145, 148
Coalbrookdale, Salop 115, 151
Coelenteron 110, 111, 115
Coelom 48
Coenenchyme 116
Coenosteum (-ea) 118–19
Coiling 37, 66–7
Collagen 2, 34, 102, 105
Colonial habit 30, 110–11
Colorado, USA 82, 122, 153
Colpus 139
Columella 69, 112, 114–15, 117
Columnal 107
Commissure 46, 49, 54, 56, 62–3, 89, 91, 97
Common canal 30
Compartmentalization of cell function 134–5
Compound eyes 18, 19, 21
Compound plate 103, 105, 107
Compressions 2
Computers 10
Concave 76, 89
Conch 72
Conchiolin 2, 89, 96
Condensate 140
Cones 142, 145–6, 150

172 *General Index*

Conifers 140–50
Conispiral 88
Connecting ring 78, 85
Connecting sheet 111
Connective tissue 53
Conodont animal 142
Conodont apparatus 137, 142
Contentin, France 88
Continental collision 163
Convergence, *see* 'convergent evolution'
Convergent evolution 3, 33, 126, 156
Convex 76, 89
Coprolites 1
Coral reefs, *see* 'reefs'
Corallite 114, 116, 117
Corallum 110–11, 113–15
Corded keel 76, 89
Correlation 5
Cortex 31, 34
Cortical tissue 34
Costa (-ae) 46, 56–8, 60, 62, 63
Cotswolds, England 105–6
Counter-cardinal 113
Counter-lateral 113
Coxa (-ae) 20
Cranidium (-ia) 18
Cranium (-ia) 123
Crenella (-ae) 107
Crocodile 126
Cross-pollination 138
Crossed lamellar structure 71, 90
Crossing canals 34
Crown 107
Cruralium (-ia) 50
Crus (crura) 56
Cryptic habitats 53
Cultural evolution 129
Cuticle 3, 20, 53
Cuttlefish 74, 79
Cylindrical 112
Cypris Freestone 142
Cyrtocone 78, 83
Cyst stage 132
Cytoplasm 134
Czechoslovakia 144

Danygraig, Swansea, Wales 151
Dating rocks 5
Declined 32
Deflexed 32
Degrees 21
Deltaic Series 150
Delthyrial cover 51, 53
Delthyrium 46–7, 50, 53, 57–8, 60, 62, 63
Deltidial plates 47, 50, 63
Deltidium 50, 62
Dendrogram 10
Dendroid habit 37, 112, 117
Depth-related communities 54
Derived fossils 6, 84
Determinism, *see* 'evolutionary determinism'
Detorsion 69
Dextral coiling 69
Diachronism 4
Diademoid 103
Diagenesis 4
Diatomite 132
Dichograptid development 34, 40
Dichotomy 144
Dicyclic 107
Diductor muscle 46, 48, 50–1, 96
Digestive gland 64, 74, 90
Digit 122, 155
Digitation 68
Dimorphism, *see* 'sexual dimorphism'
Dimyarian 92

Dinosaurs 127–8
Diplograptid development 34, 36, 40
Diploid 135–6
Dirt beds 127
Disjunct 67
Dispersal mechanisms 138
Dissepimentarium 112, 114
Dissepiments 31, 37, 112, 114–15, 117
Dissolution 4
Distal 34
Division 132
Division, of function in cells, 134, 135
DNA, *see* 'nucleic acids'
Dodo 163
Dogger Bank 98
Dogfish 120, 124
Dolphins 126, 157
Domical 118–19
Dorsal 16, 44
Dorsal furrow 18, 20
Dorset, England 104, 117
Dorsum 76
Doublure 22
Downtonian 144
Dyfed, Wales 14, 38
Dysodont 94
Duck-billed platypus 127
Dundry, Somerset 95–6
Duplicature 141

East African Rift Valley 129
East Anglia 84
Ecdysis 16, 21
Ectoderm 110
Ectoplasm 135
Ediacara, Australia 13
Edge zone 111
Egg membranes 126–7, 129
Eifel, Germany 115
Elgin, Iowa, USA 29
Endobyssate 96
Endocones 78
Endoderm 110, 115
Endoplasm 134–5
Endosiphuncular deposits 78, 85
Enrollment 17, 20
'Entire' pallial line 93–5
Epibyssate 96
Episeptal deposits 78
Epitheca 110, 112, 114
Epithelium 48, 102
Epivalve 134
Equatorial axis 139
Equatorial region 100
Erect posture 129
Escape reflex 97
Escutcheon 91, 95
Euryhaline 53
Evershot, Dorset 88
Evolute 76, 82, 88–9
Evolutionary determinism 123–4
Excrement, fossil 1
Exine 140
Exoskeleton 16, 66
Expansion, whorl 66–7
External mould 4
Extinction, *see* 'mass extinction'

Facet 20
Facial suture 18–19
Falcate 76
Falcoid 76, 89
Rall R., S. Dakota, USA 89
Family 9
Fasciole 102

Fasciculate 112–17
Faunal type 38
Feathers 127
Femur 122
Ferns 136
Fiat, motor manufacturers 155
Fibia 122
Fibrous layer 52
Fig-Tree Chert 131
Fixed cheek 18, 26
Fixigena 18, 26
Flagellum (-a) 133
Flanges 140–1
Flight 127–8
Flowers 153
Fold 46, 49
Folkestone, Kent 88
Food chain 3
Food groove 46, 107, 109
Food vacuole 110
Foot 64–5, 69, 74, 84
Foramen (-ina) 31, 36
Foramen magnum 129
Foramina 135
Ford, motor manufacturers 155
Forest of Dean, Gloucs 151
Form genera 146
Formic acid 142
Fossil 1
Fossil forest 3
Fossula (-ae) 112–14
Free cheek 18
Fringing barrier reef 118
Frustule 132, 134
Fulcrum 53
Fungi 3, 130
Funnel, *see* 'hyponome'
Fusellae 34
Fusellar half-ring 31
Fusellar layer 31, 34
Fusiform 68, 136

Gallery 118
Gametophyte 136–8, 148, 153
Gamont 135–6
Ganglia 65, 74, 90
Gape 86, 92, 95
Gas 140
Gastrovascular cavity 111
Genal angle 18
Gene pool 158–9
Generating curve 67
Generic classification 8
Geniculate 32
Genital plate 100
Genital pore 100
Genito-visceral ligament 74
Genome 161
Genus (-era) 8
Geoduck 92
Geological hammer 80
Germinal aperture 140
Germination 138
Gestation period 128
Giant panda 157
Gilfach Goch, Wales 151
Gill arches 122, 124
Gill branches 17
Gill slits 122–3
Gills 64–5, 67, 72, 74, 84, 86, 90, 101, 122, 124
Girdles 121
Glabella 18–19
Glabellar furrow 18
Glabellar lobe 18
Glabellar segment 18
Glaciation 6, 139–40

God 158
Gonatoparian 19, 22
Goniatites 74
Goniatitic suture 77, 86
Gotland, Sweden 120
Gower, Wales 117
Gradualism 40
Grass 152
Gravity, centre of, 80
Great Barrier Reef, Australia 118
Greenland 126
Growth line 46, 50, 52, 56–7, 62, 71, 74, 75, 90, 93–4, 96–7
Guard 78–80
Gwynant Valley, Dolgellau, Wales 37
Gymnosperms 138
Gyres 39

Haematite 4
Haemocoel 74
Haploid 135–6
Harz Mts, Germany 108
Hatchback motor-cars 155
Haverfordwest, Wales 58
Head-foot, *see* 'foot'
Head-shields 122–3
Heart urchin 103
Heath Robinson 156
Hedgehogs 129
Helical coiling 66, 73
Helter-skelter 66
Herbury, Dorset 63
Herefordshire, England 144
Hermaphroditism 44, 69, 87
Hermatypic 116
Heterodont 93
Heteromorph ammonites 81–2, 162
Heterospory 137–8, 144, 148
Hexactinellid network 119
Hinge 44, 51–2, 84, 90
Hinge axis 51–2, 91
Hinge line 45, 51, 56, 58, 94–5
Hinge plate 56, 91, 93
Hip-girth 82
Holaspis 21, 29
Holdfast 35, 37
Holostomatous 70
Homeostasis 13
Hominid evolution 129
Homoeomorphy 156–7
Homology 129, 155
Homospory 137–8, 148
Hood 74
Horn shape 98
Horse 157, 161
Horsetails 146
Humerus 122, 155
Hybrids 8
Hydrocarbons 140, 143
Hydrodynamics 79–80
Hydrofluoric acid 140
Hydrostatics 79–80
Hyomandibular 124
Hyponome 66, 74–5, 77
Hyponomic sinus 77–8
Hyposeptal deposits 78
Hypostome 22
Hypovalve 135

Iapetus 28–9
Immune system 129
Immunosuppressants 129
Impregnation 3
Impressions 2
Incisors 127
Incompatibility, *see* 'reproductive incompatibility'

174 General Index

Incurved whorl   88
Index fossil   5
Inequivalve   93
Infrabasal plates   107–8
Infraradial plates   107
Inhalent siphon   69
Ink sac   75
In situ   5
Indiana, USA   108
Instar   16
Interarea   46, 51, 58, 60, 62–3
Interambulacrum (-cra)   100–2, 104–6
Interbrachial plate   109
Interbreeding   8
Interglacial sediments   139
Internal lamella   140
Internal mould   4, 75, 78, 85–6, 88
Internal shell   75, 78–9
Internal skeleton   99
International Code of Zoological Nomenclature   10
Interpleural groove   20–1
Inter-radial plate   109
Intine   140
Introtorted   32
Introverted   32
Invertebrates   2
Involute   67, 71, 76, 82, 87–8
Irish Elk,   3
Ironstones, Precambrian,   13
Irreversibility of evolution   71, 87
Isolate   32
Isomyarian   90, 92–3, 95, 97
Isopygous   21
Iterative evolution   84

Jawbone   123
Jawed vertebrates   121
Jet propulsion   75, 81
Jones Co., N. Carolina, USA   104

Karroo Beds   127
Keel   76, 78, 88–9
Keratin   2
Keyhole limpet   68
Kingdom   9, 130
Kneippbyn, Gotland, Sweden   120

Labial palps, *see* 'palps'
Labrum   104, 160
Laggan Burn, Scotland   40
Lagoon   118
Lake District, England   39, 41
Lamp-shells   44
Lantern, *see* 'Aristotle's Lantern'
Lappets   77, 82–3, 89
Larva   21
Lateral fins   122–3
Lateral teeth   91
Latex replicas   4
Latilamina (-ae)   118–19
Leaf scars   146
Leaves, origin   148–9
Leintwardine, Salop   42
Leprechauns   131
Leptograptid development   34
Librigena   26
Ligament pit   91, 93–4, 96–7
Lightning   12
Limonite   4
Limpet   66–7
Listrium   47, 55
Litopterns   151, 161
Llangollen, Wales   57, 114–15
Lobe   77, 86–7

Lobe-fins   121, 125
Lobed thecae   32
London Clay   161
Looped ribs   76, 89
Lophophore   45, 48, 54
Lucinoid bivalves   98
Lucinoid dentition   93
Ludlow, Salop   23, 55
Lungfish   121, 124
Lumbago   129
Lumen   107
Lung   68
Lunule   91, 95
Lycopods   141

Mackerel   80
Macrocanals   105
Macroconch   82–3
Macropygous   21
Madreporite   100–1, 106
Magnetite   105
Mamelon   102, 118–19
'Mamelon & boss' tubercle   105
Mammal-like reptiles   127
Mammary glands   127
Mammoth   3
Manchester, Lancs   159
Mandibular palp   141
Manhattan, New York, USA   98
Mantle   45, 48, 51, 64, 84, 89
Mantle caecum   48
Mantle cavity   45, 48, 64–5, 67–9, 75, 84, 87, 95
Mantle flaps   68
Mantle fold   52
Mantle groove   48
Marginal facial suture   19
Marginal fold, *see* 'edge zone'
Marginal furrow   18
Marginal spines   21
Marine band   83
Marker band   98
Marker fossil   5
Mark's Tey, Essex   139–140
Marsupium   128
Mass extinction   61, 83–4, 106, 127, 161, 163
Massive   112
Maturation   140
Maxilla (-ae)   141
Meckel's Cartilage   124
Median septum   56, 58
Median suture   102, 105–6
Megaspore   131, 137–8
Meiosis   136
Melanism   159
Meraspis   21, 29
Mesentery   110–11
Mesogloea   110
Metacarpal   122, 155
Metaseptum,   110, 113, 114
Metasicula (-lae)   31
Metatarsal   122
Micrite   120
Microcanals   105
Microconch   82–3
Micrometre (μm)   131
Micropalaeontology   1
Micropygous   21
Microsporangium (-ia)   148
Microspores   131, 137–9
Mid Atlantic Ridge   163
Milioline winding   136
Mineralized chitin   20
Missing links   162
Mitosis   136
Moffat, Scotland   39, 40, 43
Molar teeth   127

Monocyclic 107–8
Monograptid development 34, 36
Monkey Puzzle Tree 150
Monolete 139–40
Monomyarian 90, 92, 96
Monopodial 144
Monosulcate 139–40
Monotremes 127
Morphospecies 9, 82
Mosses 136
Mother-of-pearl 71
Moulds and casts 4
Much Wenlock, Salop 62, 115
Mucoprotein 48
Mucus 84, 96, 98
Mule 8
Mull, Isle of, Scotland 150
Multi-storey car-park 118–19
Mummification 3
Mural deposits 78
Mural pores 113
Muscle scars 46–8, 50, 56, 58, 84, 90–2, 141
Mutation 159
Myophore 95, 97

Naticiform 70
Natural selection 158–9
Neck 70, 122
Nektic organisms 6, 16
Nema 30–1
Neoteny 97, 129
Nephridiopore 46, 65
Nephridium (-ia) 46, 48, 65
Nerve cord 121, 129
Neutral buoyancy 79
Newcastle, Northumbria 150
Node 76, 87–9, 147
Non-marine bivalves 98
Non-strophic 46, 51–2, 56, 62–3
Norfolk, England 109
North Sea 98
Nostril 122, 124
Notochord 121–2
Notothyrium 47, 50, 53, 56
Nova Scotia, Canada 72–3
Nucleic acids 13
Numerical taxonomy 10
Nummulitic Limestone (Eocene) 135
Nursing habit 121, 127

Oblique muscles 46–7
Occipital furrow 18
Occipital glabellar segment 18
Occipital node 18
Octopus 74
Ocular plate 100–1, 106
Ocular pore 100–1
Ocular ridge 19
Oesophagus 64
Oil industry 5, 84, 118, 140
Oklahoma, USA 40, 85, 88
Olduvai Gorge, Tanzania 129
Ommatidium (-ia) 18
Onlap 6
Ontogeny 21, 113
Ooze 134–5
Operculum 68–9, 77, 115
Opisthodetic 94
Opisthogyrate 90, 92
Opisthoparian 19, 22
Opposable thumb 129
Oral 99–100
Orbit 124
Orciano, Italy 93
Order 9

Orthocone 78, 80–1, 83, 85
Orthogenesis 161–2
Osphradium (-ia) 64
Ossicle 107
Outbreeding 153
Outer lamella 141
Outer ramus 22
Ovule 138, 152
Oxidation of pyrites 4
Oxycone 76, 86
Oxygen isotopes 6
Oxygen, origin 13
Oyster 92, 98
Ozone 13

Pacific Ocean 118
Pacific Province 29
Pakistan 129
Palaeobiology 1
Palaeobotany 1, 2
Palaeoecology 1
Palaeontology 1
Palaeontological Association, The 10
Palaeotaxodont 93
Palaeotemperature 6, 133
Palato-pterygo-quadrate bar 124
Palisade 114–15
Pallial line 91–5
Pallial markings 56
Pallial sinus 91–4, 97
Pallium 64, 92;
See also 'mantle'
Palms 152
Palp 86, 90, 141
Palpebral lobe 19
Palpebral furrow 19
Palus 117
Palynology 1, 140
Pangaea 163
Paradigm 154–7
Parietal callus 70
Patellate 112
Peat bogs 2, 3
Pectoral fin 123–4
Pectoral girdle 122
Pedal ganglion, see 'ganglia'
Pedal gape 97
Pedal retractor muscles 84
Pedicellaria (-ae) 101–2
Pedicle 44, 46–7, 50–1, 53
Pedicle foramen 47–8, 50, 56, 62
Pedicle valve 44, 46–7, 88
Pedunculate 44
'Pelecypods' 84
Pellets 64
Pelvic girdle 122, 129
Pembrokeshire, Wales 144
Penarth, Wales 96
Penis 68
Pennate 132, 134
Pennine Alps 84
Pentadactyl limb 122, 154–5
Pentameral symmetry 99
Pericardium 65, 90
Periderm 31, 34–5
Perignathic girdle 101
Periostracum 48, 52, 69, 90, 96
Periproct 100, 105–6
Peristome 76, 78, 101, 105
Permineralization 4
Petaloid ambulacra, see 'petals'
Petals 101–2, 104
Petri Dish 132
Petrifactions 2
Petroleum 3
Phaceloid 112, 114

Pharynx 111
Phenetic groups 22
Phosphorus 116
Photic zone 116, 132
Photosynthesis 13, 115–16, 130–2
Phragmocone 75–6, 78–80, 86
Phycocyanin 131
Phyletic extinction 163
Phyletic gradualism 162
Phylogenetic classification 22, 89
Phylum 9, 132
Pillar 118–19
Pineal eye 123
Pinnae 148–9, 151
Pinnule 107–8, 148–9
Pith cast 147
Pith cavity 147
Pits 25
Placenta 129
Placental mammals 128
Plankton 6
Planispiral 67, 71, 76, 81, 83, 86, 136
Planulate 76
Plastron 101–2, 104, 106
Platforms 47, 49, 51, 53, 56, 142
Pleural furrow 20–1
Pleuron (-a) 16
Plication 46, 49, 54, 56, 60, 62–3, 94, 98
Podia 101
Podolia USSR 144
Poison glands 101
Polar axis 139
Polar nodule 134
Pollen 130, 135, 138–9, 150
Pollen analysis 139–40, 153
Pollen assemblage 139–40
Pollen tube 138
Polygyrate 76
Polymorph 2, 4
Polyp 110, 114–15
Polypeptides 13
Polyphyletic 33, 84
Pongola System 131
Pore canals 141
Pore pairs 100, 103–6
Porous plate 100
Posterior 16
Pouch 128
Powered flight 128, 162
Preadaptation 70, 124, 162–3
Pre-ferns 148
Pregnancy 128
Preservation potential 2, 14
Preservational eclipse 127
Primary septum 112–13
Primary shell 48, 52
Primates 9, 129
Primordial atmosphere 12
Principal spines 134
Problematica 35
Proboscis 36, 68, 72
Pro-ostracum 78–9
Proparian 19, 22
Prorsiradiate 76
Proseptum 110, 113–14
Prosicula 31
Prosiphonate 78
Prosogyrate 90, 93
Protaspis 21, 29
'Proteinoids' 13
Proteins 12
'Protobrachiopods' 60
Protoconch 74, 78–9, 83
Protoplasm 134
Provincialism 26, 29, 39
Proximal 34
Proximate 133

Pseudopodia 135
Pseudopunctate 58–9
Pterobranch 36
Pteropod ooze 72
Puncta (-ae) 48, 52, 56, 134
Punctuated equilibrium 40, 162
Pupiform 67, 70
Puryear, Tennessee, USA 2
Pygidium 17, 21, 65
Pyrites 3

Quadrate 86, 89
Quadriramous 33
Quadriserial 33
Quick adductor 52

Rabbit 127
Radial canal 103
Radial plates 107
Radial symmetry 99
Radial sesamoid 157
Radiometric dating 5
Radius 122, 155
Radstock, Somerset 96
Radula 64, 66–7, 72
Ramus 22
Raphe 134
Razor clam 91–4
Reclined 32
Recrystallization 4
Rectiradiate 76, 88, 89
'Red tide' 132
Reefs 53, 61, 71, 84, 88, 92, 102, 113–15, 117–18, 120
'Reef rock' 120
Reflexed 32
Relative dating 5
Reniform 23
Reproductive incompatibility 159
Reptile-like mammals 127
Respiratory skin 126
Respiratory tube-feet 104, 160
Retrosiphonate 78, 85
Rhabdom 18
Rhabdosome 30
Rhinoceros 3
Rhizome 144, 147–8
Rhydymwyn, Wales 56
Rhynie Chert 144
Ribs 56, 76, 78, 80, 87–90, 93–4, 141
Ring canal 101
Robin Hood's Bay, Yorks. 3
Rocket engine 82
Rostrum 103–4
Rubber balloon 99
Rudist bivalves 88, 92, 98
Ruga (-ae) 49, 58
Rugose epitheca 115
Rursiradiate

Saccate 140
Saccus 139
Saddle 77, 86–7
Salisbury, Wilts 88
Sand dollar 102, 104–5
Sanitary drain 102–3
Scaffolding 118–19
Scallop 92
Scandent 32–3
Schistes Lustrés 84
Schizont 135–6
Scleroprotein 2, 34
Sclerosepta 117
Scolecoid 112
Scolecodonts 130

Scrobicular tubercle 102, 105
Sea-floor spreading 135
Sealed egg 126–7
Secondary growth 146
Secondary septum 112–13
Secondary shell 48, 52
Secondary sexual characteristics 83
Secondary tubercles 105
Sedimentology 118
Seed 138, 144–53
Seed-ferns 148
Seed-leaves 153
Segment 17, 20–2, 25–7, 28–9
Segmentation 65
Selenizone 68–9, 71–2
Selenopeltid Fauna 29
Selvage 141
Septal insertion 113–14
Septal neck 74, 78–9
Septal traces 115, 117
Septum (-ta) 50, 72, 74–5, 78, 85, 110–12, 136, 155–6
Serial sections 113
Serpenticone 76, 89
Sessile 87
Setae 25, 47–8, 50, 54, 141
Sex-ratio 83, 141
Sexual dimorphism 31, 75, 82–3, 141–2
Sexual display 83
Sharks 2, 121, 124, 157
Sharks' teeth 2, 121
Shelly facies 26
Shelve, Salop 37
Shipworm 92, 95
Shoulder 71
Shrews 129
Siberia, USSR 3
Sicula 30–1
Siderite 4
Sieve membrane 132
Sigmoid theca 32
Silica 2, 132
Sinistral coiling 69–70
Sinus 91;
See also 'pallial sinus'
Siphon 54, 64–6, 68, 84, 88, 92
Siphonal canal 68, 70, 72
Siphonostomatous 68, 70
Siphuncle 74–5, 83, 85
Skelgill, Lake District 39, 42
Skull 121–2
Slipper limpet 72
Slit-band, see 'selenizone'
Sloths 3
Slugs 67
Snorkel 66, 70, 84, 104
Sockets 44, 48, 50–1, 56, 58, 91, 93, 140
Solar system 8
Solnhofen Limestone 128
Solution pipe 127
South Africa 127, 131
South Dakota, USA 89
Spatulate 23, 55
Species 8
Spermatozoon, largest 140
Sphaerocone 76
Spicule 2, 130
'Spider's web' 8, 9, 112–14
Spiegeltown, New York, USA 39
Spiral brachidia 49–50, 52
Spire 70
Spired echinoids 102
Spinal column 121
Spine-base 102, 105
Spinnerets 3
Splint-bones 161
Spondylium 50
Sporangium (-a) 144, 148

Spore 3, 130–1, 135
Spore-assemblage (Devonian) 148
Sporophyll 145
Sporophyte
Spreading centres 163
Squid 74
Stabilizers (ship's) 80
Stalked eyes (trilobite) 19, 27
Statocyst 64
Steeple Ashton, Wilts 117
Steeple Hill, Wilts 117
Steinkern 4
Stem 106
Stenohaline 53, 106
Stigmarian base 146
Stings 72
Stipe 30–1
Stolon 31
Stolotheca 31
Stone Age 129
Stone canal 101
Stonesfield Slate 127
Stratigraphy 4
Stria (-ae) 31, 134, 141
Stromatolites 131–2
Strophic 46, 51–2, 56–8
Sub-anal fasciole 102, 160
Subfossils 71
Submarine 75, 79–80
Suctorial mouth 123
Sulcus 46, 49, 60, 62–3, 76, 78, 106, 133
Supercolony 36–7
Supercontinent 163
'Superphylum' 9
Superposition, Law of 5
Supraradial plates 107
Sutural ramp 70–1
Suture 31, 70, 75, 77, 79–80, 83, 85, 86–8, 89, 101, 136, 155–6
Swim bladder 125
Symbiosis 98, 115, 132
Synrhabdosome 34, 36–7
Systema Naturae 9

Tabula (-ae) 110, 112, 114–15
Tadpole 126
Tail 122
Tar-pit 2
Tarsal 122
Taxodont 93, 95
Taxometrics 11
Taxonomy 8
Tectin 135
Teeth 2, 44, 48–9, 51, 60, 91, 93–7, 121, 140
Tegmen 109
Telome concept 149
Telopodite 22
Telson 21
Tentacles 36, 66, 68, 74, 75, 111
Terminal bud 36
Terminal sporangium, *see* 'sporangium'
Terrace 102, 105
Test 99, 135, 159
Tethyan Realm 98
Tetrad 139
Tetragonal tetrad 139
Tetrahedral tetrad 139
Tetrapod 120
Thames River, England 6
Theca (-ae) 30, 111, 133
Thecal angle 32
Thecal aperture 32
Thecal length 32
Thecal morphology 32
Thorax 17, 65
Thorns 3

Three-dimensional vision 129
Tibia 122
Time-contour 5
Tollund Fen, Denmark 3
Tollund Man, 3
Tongwynlais, Wales 149
Torsion 65, 68–9
Trace fossil 1
Tracheophytes 144
Tracheids 3
Trail 49, 54, 57–8
Transgression 5
Translation, of whorl, 66–7
Transplant surgery 129
Transported fossils 5
Treatise on Invertebrate Paleontology 22
Tree-ferns 148
Trema (-ata) 68–9, 71
Tricarinate 76
Trichotomy 144
Tricolpate pollen 139, 140
Trifurcate 76
Trilete mark 139–40, 144
Trilete spores 140
Trinucleid brim 25, 27
Triserial 136
Trochiform 70
Trochoid 112, 115
Trochospiral 136
True ferns 148
Tube-feet 99, 101, 103
Tubercles 19, 78, 80, 88, 93, 100–1, 103–5, 141
Tubiform 71
Tufa 2
'Tuning fork' graptolite 38
Turbinate 70
Turreted 70
Turriculate 70
Turtle 70
Type specimen 10

Ulna 122, 155
Ultraviolet light 13, 135
Umbilical seam 76
Umbilicus 76, 86–8, 136
Umbo 46, 90
Uncoiled gastropods, see 'vermiform gastropods'
Unconformity 6, 52
Unicarinate 76, 89
Uniformity, Principle of 6
Unilocular 136
Uniserial 33
Urine 46
Uterus 129
Utica Shale 20

Vacuole 110, 135
Valves 44, 66, 84, 140–1
Varicose veins 129
Vascular bundle 147
Vascular plants 144
Vauxhall, motor manufacturers 155
Venter 76, 86–9
Ventral 16, 44
Vermiform gastropods 70, 72
Vertebra (-ae) 121
Vertebral column 121–2, 124
Viola Limestone 40
Virgatotome 76
Virgella 32
Virgula 36
Visceral ganglion, see 'ganglia'
Vivipary 127
Volcanic sediments 1, 3, 129
Volcanoes, and origin of life 19
Volume-to-surface ratio 131

Wales 83, 92, 127
Warm-blooded dinosaurs 127
Warren's Truss 128
Water-vascular system 99–101
Wealden Series 127, 150
Weighted characters 10
Wenlock Edge, Salop 116
Wenlock, Salop 26;
See also 'Much Wenlock'
Whelk 68
Whitby, Yorkshire 77, 88, 152
Whorl 66, 76, 86–7
Wiltshire, England 88–9
'Winged oyster' 92
'Wings' (brachiopod) 60, 62
Wolstonian 139
Wrexham, Wales 55

X-rays 20

Year, Devonian 7–8
Yellowstone National Park, USA 3
Yolk 126

Zig-zag 54, 56, 63, 80, 101
Zonation, ecological 53, 118, 120
Zone 5, 83, 106, 109, 117, 134
Zooid 30, 110, 115
Zooxanthellae 115–16, 120, 132